Holmes McDougall Metric Maths

BOOK 4

J. O'Neill
J. Potts
J. Wood

General Editor: I. D. Watt

Holmes McDougall Ltd

A

Art-work by F. Vaughan
Printed by Holmes McDougall Limited

SBN 7157 0783-3

Contents Book 4

INTRODUCTORY NOTE

The whole atmosphere of education has changed in the post-war years, particularly in the last decade. Everywhere, at all stages, there is evident a quickening of interest and a lively urge to try out new ideas and new methods, to provide an education that makes sense to the children of to-day. Teachers are aware, as no-one else, that there is no room for complacency, that we are nearer the beginning than the end of educational development. What can be claimed is that the prospect of sensible and sustained advance is brighter than ever before.

Nowhere is change more evident than in the Primary School. The day of the stereotype in syllabuses and textbooks is quickly going. New subjects have been introduced, old ones refurbished and modernised. Teaching methods are placing greater emphasis on assignments and pupil participation. The stress is on adaptability and the capacity to think creatively. The demand for change has come from the schools themselves. The practising teacher in the classroom is becoming more and more the pacemaker and the initiator of new ideas of what to teach and how to teach.

Of all subjects in the Primary School curriculum, mathematics is the one that is undergoing the greatest change. Up until a few years ago the teaching of the subject was in large part confined to the development of skill in reckoning and the manipulation of unwieldy computations of little relevance and doubtful educational value. Much of the work was of a routine nature, much of it divorced from practical realities, with little understanding of the nature of the underlying concepts. This is being changed.

Materials for the study of mathematics—number, quantity and shape—abound in the child's environment and, by means of meaningful experiences, purposefully planned, children are being led progressively to the understanding of basic mathematical concepts. But, as the Memorandum on Primary Education in Scotland points out, there is no suggestion that all the material to be studied can be found directly in the environment or that none of the subjects grouped under Environmental Studies has to be pursued as a separate discipline. In the words of the Memorandum: "Indeed, in the case of mathematics, especially that part of it which is concerned with number, there must be from the earliest stages training in specific skills that will enable the pupils to handle quickly and efficiently the mathematical situations which will arise in the course of their other activities."

In this series of development exercises it is the need for this basic training in specific skills that the authors—all experienced in the day-to-day work of the schools—have had constantly in mind. Drafts of the exercises were tried out in a variety of schools and modified in the light of the reactions of both teachers and pupils. The result is a series of books which, it is confidently expected, will make a distinctive and useful contribution to the teaching of mathematics in primary schools.

Reading and Writing Numbers

MILLIONS

ONE MILLION is written 1 000 000.
FIVE MILLION is written 5 000 000.

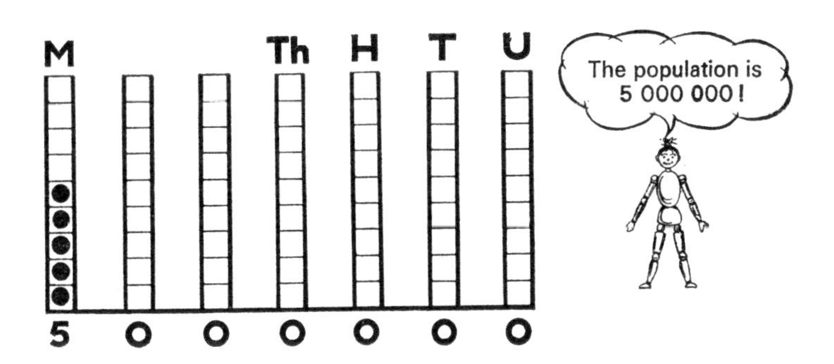

THIRTY-NINE MILLION is written as 39 000 000.
253 649 178 is read as, "two hundred and fifty-three million six hundred and forty-nine thousand one hundred and seventy-eight".

Note: After the number of millions, there are *six* digits.

(1) Copy and complete this table:

	WRITE	READ
(a)	7 000 000	seven million
(b)		five million
(c)	10 000 000	
(d)		eighteen million
(e)	23 000 000	
(f)		ninety-two million
(g)	150 000 000	
(h)		two hundred and thirty-five million

(2) Read these numbers and write in words the value of the 3 in each:

(a) 8 000 030 (b) 6 030 000 (c) 7 253 000 (d) 2 536 000
(e) 9 127 300 (f) 4 382 760 (g) 3 781 496 (h) 8 036 147
(i) 1 205 310 (j) 4 380 065 (k) 7 300 206 (l) 5 003 007

USING LARGE NUMBERS

Abe says that the sun is about one hundred and forty-eight million, eight hundred thousand kilometres away from the earth.

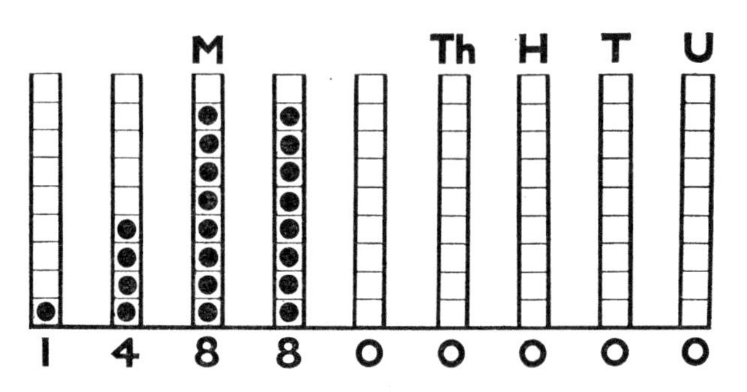

In figures: **148 800 000**

2 *More Large Quantities:* Write in figures the large numbers in the following:

(1) Venus is about one hundred and seven million, two hundred thousand kilometres from the sun.

(2) Mars is about two hundred and fifteen million, six hundred thousand kilometres from the sun.

(3) The moon is about three hundred and eighty-two thousand, two hundred and forty kilometres from the earth.

(4) A bullet may travel at two thousand, two hundred and forty kilometres/hour.

(5) A satellite travels at twenty-seven thousand, two hundred kilometres/hour.

(6) Light travels at two hundred and ninety-seven thousand, six hundred kilometres/second.

(7) The population of China is about six hundred and seventy million people.

(8) The population of U.S.A. is about one hundred and seventy-nine million, three hundred and twenty-three thousand people.

Write in figures the answers to these:

(9) 999 999 + 1
(10) 999 999 + 10
(11) 999 999 + 100
(12) 999 999 + 1000
(13) 1 000 000 1
(14) 1 000 000 − 10
(15) 1 000 000 − 100
(16) 1 000 000 − 1000

CAN YOU READ AN ELECTRICITY METER?

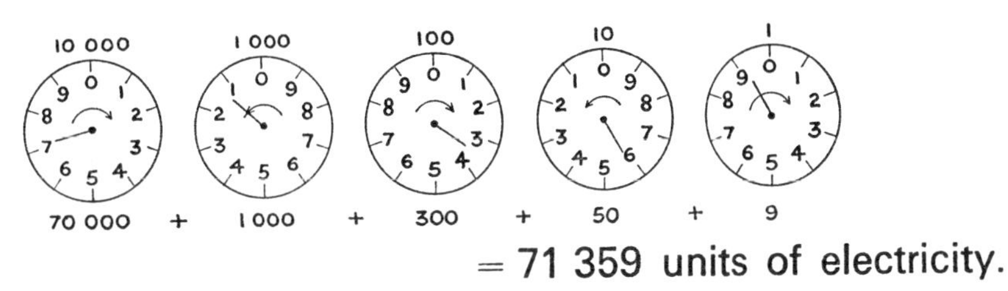

= 71 359 units of electricity.

Note: Each dial supplies one digit for the number of units of electricity.
The value of each digit depends on the position of the hand on the dial.

3 Read these electricity meter dials and write down your "readings":

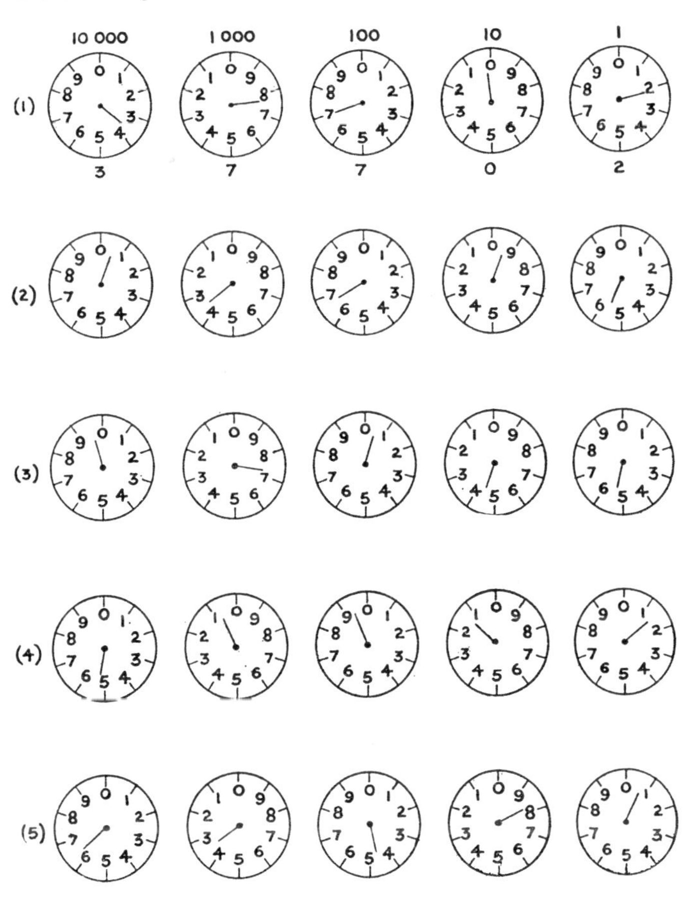

Addition and Subtraction 

PRACTICE

Write down the answers to:

4

(1) 4 + 7 (2) 5 + 9 (3) 3 + 8 (4) 8 + 9 (5) 4+7+9
(6) 12 − 6 (7) 15 − 7 (8) 14 − 7 (9) 16 − 7 (10) 20 − 5
(11) (8 + 7) − 3 (12) 20 − (9 + 4) (13) 18 + (6 − 3)
(14) 16 − (7 + 2) (15) (9 + 8) − 7 (16) 14 + 12 + 6
(17) 38 + 21 − 8 (18) 33 − 16 + 5 (19) 37 − 18 + 13
(20) 42 − 11 − 12

Add:

5

(1)	(2)	(3)	(4)
346	72·4	218	3·76
529	38·5	329	8·93
183	9·6	571	6·51

(5)	(6)	(7)	(8)
20·38	7 243	2·218	6 804
9·53	3 495	0·374	407
11·22	4 371	1·56	3 593

(9)	(10)	(11)	(12)
51 306	26·35	36 702	18·642
5 432	318·92	52 816	9·879
20 645	273·56	29 753	21·346
21 716	18·94	34 149	49·23

(13)	(14)	(15)	(16)
354 951	7 230·15	1 038·32	902 081
114 420	6 028·10	4 223·25	91 909
543 025	2 466·62	2 544·03	83 837
872 263	3 832·12	9 107·16	710 094

Subtract:

6

(1)	(2)	(3)	(4)
542	98·6	714	2·33
329	29·3	165	0·46

(5)	(6)	(7)	(8)
75·38	6 807	71·65	5 348
32·43	945	22·94	1 729

(9)	(10)	(11)	(12)
36 975	13·736	290 623	233 715
24 138	4·581	244 834	161 692

1 823
+3 594
5 417
−2 067
3 350

1823 + 3594 − 2067

Find the answers to:

7

(1) 548 + 356 − 231
(2) 735 − 148 + 207
(3) 323 + 189 − 84
(4) 901 − 256 + 187
(5) 6593 − 2626 + 347
(6) 3893 + 3651 − 2584
(7) 6435 − 2583 + 5269
(8) 5278 + 3756 − 716
(9) 5679 − 4087 + 5885
(10) 10 000 − (425 + 1293)

PROBLEMS

8

(1) The attendance on a certain morning at Central School is shown on the board:
(a) How many pupils were on the roll?
(b) How many pupils were present?
(c) How many pupils were absent?

Class	Roll	Present
1	38	34
2	42	40
3	39	32
4	43	36
5	39	37
6	40	38
7	34	29

(2) Subtract the smallest of the following numbers from the largest: 2 065, 3 892, 4 231 and 2 203.

(3) 46 834 tickets were sold for the Enclosure at a football match and 5 276 tickets were sold for the Stand. How many tickets were sold altogether?

(4) 86 300 tickets were printed for an international hockey match. 52 635 tickets were sold. How many tickets were left?

(5) A television set was on sale in the shop at £73·50. Dad bought the set and had to pay an extra £6·75 for the legs and £3·40 for an aerial. How much did Dad spend altogether?

(6) A playing field is in the shape of a rectangle 216 metres long and 94 metres broad. What length of fencing is needed to enclose the field?

(7) At an election the votes cast for the candidates were as follows:

James Adams	18 346
Ian Blair	12 471
John Clark	4 588

(a) How many votes more than Mr Blair did Mr Adams receive?

(b) How many votes more than Mr Blair and Mr Clark together did Mr Adams receive?

Time — The 24-Hour Clock

ONE DAY IS DIVIDED INTO 24 HOURS

When writing times on the 24-hour clock we use four digits:
two for the hour
and two for the minutes past the hour.
We count the time from midnight. For example 2 a.m. = 02 00

Abe's Clock

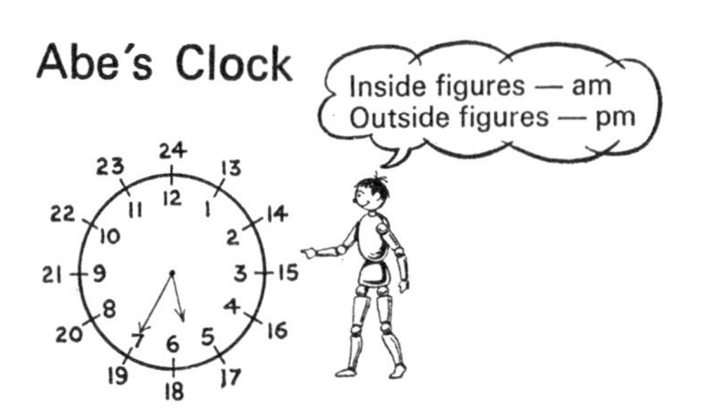

DIGITAL CLOCK

17 35

This clock has no hands.

It shows the digits for the hour and for the minutes past the hour.

Both clocks are showing the same afternoon time:
17 35 or 5.35 pm
= 35 minutes past 5 in the evening
or 25 minutes to 6 in the evening

Abe's clock could also be read as:
05 35 or 5.35 am
= 35 minutes past 5 in the morning
or 25 minutes to 6 in the morning

Write the following 24-hour clock times using am and pm:

9

(1) 03 45 (2) 19 27 (3) 21 45 (4) 07 52

(5) 09 15 (6) 20 30 (7) 10 02 (8) 15 05

(9) 03 10 (10) 05 05 (11) 00 15 (12) 21 20

(13) 12 35 (14) 23 59 (15) 00 03 (16) 10 01

(17) 22 10 (18) 04 40 (19) 13 30 (20) 23 25

10 Write the following times as they would appear on a 24-hour clock:

(1) 5.25 pm (2) 5.45 am (3) 11.10 am (4) 9.05 am
(5) 9.10 pm (6) 1.03 pm (7) 9.50 pm (8) 8.05 pm
(9) 25 minutes past 4 in the morning.
(10) a quarter past 12 mid-day.
(11) half past 12 midnight. (12) 20 minutes to five in the afternoon.
(13) 5 minutes to 6 in the morning. (14) 5 minutes past 11 at night.
(15) half past 6 in the evening. (16) 1 o'clock in the morning.
(17) a quarter to 3 in the afternoon. (18) 10 minutes to 10 at night.
(19) 20 minutes past 8 in the evening. (20) 1 minute to midnight.

TIMETABLES

London to Jersey:

Outward from London on Friday night May 15 / Sept 25

Return from Jersey on Saturday night May 16 / Sept 26

Outward			Return
20 37	dep London (Waterloo)	arr	09 04
21 03	dep Woking	arr	08 33
21 24	dep Basingstoke	arr	08 12
22 01	dep Southampton	arr	07 39
22 37	dep Bournemouth	arr	07 01
23 46	arr Weymouth Quay	dep	05 50
00 30	dep Weymouth Quay	arr	05 00
06 00	arr Jersey	dep	21 45

11 Answer the following questions from this British Railways London to Jersey timetable:

(1) Give the time of departure of the train from London using a.m. or p.m.

(2) How much is this time before or after 25 minutes to 9?

(3) How many hours and minutes is the *train* journey from London?

(4) If you are going to Jersey, how long is the waiting time at Weymouth?

Is this waiting time: (a) in the morning, (b) in the afternoon, (c) in the evening, (d) at night, (e) at midnight or (f) at mid-day? Write the best answer.

(5) (a) Which sea crossing takes longer?

(b) How much longer does this crossing take?

(c) What is the total time for the return journey to London?

Sets $\{○, △, □\}$

REMEMBER: $\{2, 4, 6, 8, 10\}$ is the set of the first five even numbers. The members of a set are listed between curly brackets and are separated from each other by a comma.

We say, "4 is a member of the set of the first five even numbers".

We write, $4 \in \{2, 4, 6, 8, 10\}$

$\in$ means *'is a member of'*. $\notin$ means *'is not a member of'*.

12

List the following sets using curly brackets:

(1) The set of odd numbers less than 11.
(2) The set of colours in traffic lights.
(3) The set of numbers less than 21 which are divisible by 3.
(4) A set of three christian names for girls which start with M.
(5) A set of three christian names for boys which start with J.
(6) The set of seasons in a year. (7) The set of days in a week.
(8) The set of months which have exactly 30 days.
(9) A set of five wild animals.
(10) A set of six games which are played with a ball.

13

State which of the following mathematical sentences are True and which are False:

(1) $9 \in \{3, 6, 9, 12\}$
(2) $28 \notin \{7, 14, 21, 28, 35\}$
(3) $20 \in \{6, 12, 18, 24, 30\}$
(4) $33 \notin \{5, 11, 18, 24, 30\}$
(5) $t \in \{u, v, w, x, y, z\}$
(6) $p \in \{m, n, o, p, q, r\}$
(7) $s \notin \{a, e, i, o, u\}$
(8) $o \notin \{s, c, h, o, l\}$
(9) $45 \in \{9, 18, 27, 36, 45, 54\}$
(10) $63 \notin \{21, 42, 84, 105\}$
(11) D $\in$ {Dee, Don}
(12) t $\in$ {cocoa, sugar, tea}

A SET MAY BE REPRESENTED BY A CAPITAL LETTER

$F = \{4, 8, 12, 16, 20, 24, 28, 32, 36\}$

can be read as 'F is the set containing the first nine multiples of 4'

or 'F is the set with members 4, 8, 12, 16, 20, 24, 28, 32 and 36'

For $8 \in \{4, 8, 12, 16, 20, 24, 28, 32, 36\}$
We may write $8 \in F$.

For $11 \notin \{4, 8, 12, 16, 20, 24, 28, 32, 36\}$
We may write $11 \notin F$.

Describe the following mathematical sentences in words:

14

(1) V = {a, e, i, o, u}
(2) B = {Alec, Abe, Arthur, Andrew}
(3) A = {a, b, c, d, e, f, g, h, i, j, k, l, m, n, o, p, q, r, s, t, u, v, w, x, y, z}
(4) T = {2, 4, 6, 8, 10, 12, 14, 16}
(5) E = {8, 16, 24, 32, 40, 48, 56}
(6) M = {July, August, September, October, November, December}
(7) F = {5, 10, 15, 20, 25, 30, 35, 40}
(8) S = {club, diamond, heart, spade}
(9) N = {1, 2, 3, 4, 5, 6, 7, 8, 9, 10}
(10) C = {$\frac{1}{2}$p, 1p, 2p, 5p, 10p, 50p}

EQUAL SETS

Equal sets are sets which have the same members

Study the sets P, Q and R

P = {p, e, t, s} Q = {s, t, e, p} R = {a, o, g, s}

The sets P and Q are equal sets since they have the same members e, p, s and t.

We may write, P = Q.

The sets P and R are not equal sets since they do not have the same members.

We may write, P ≠ R

Note: The order in which the members of a set are listed is not important.

State which of these pairs of sets are equal and which are not equal:

15

(1) A = {1, 4, 9} B = {9, 4, 1}
(2) C = {1, 2, 3, 4} D = {0, 1, 2, 3, 4}
(3) E = {10, 30, 50, 40, 20} F = {50, 60, 20, 30, 10}
(4) G is the set of odd numbers less than 8. H = {1, 3, 5, 7}
(5) I = {a, b, c} J = {1, 2, 3}
(6) K = {7} L = {7}
(7) M = {January, February} N is the set of the first 2 months of the year
(8) O = {Tom, Dick, Jean, Jane} P = {Dick, Jean, Mary, Tom}
(9) Q = {Jill, Jack, Jean, John, Joan} R = {Jill, Joan, Jean, John}
(10) S = {2, 4, 6, 8, 10} T is the set of the first five even numbers
(11) U = {cow, bull, hen, pig} V = {bull, pig, cow, hen}
(12) W = {654, 321} X = {1, 2, 3, 4, 5, 6}

AND SO ON . . .

E={2, 4, 6, 8, . . .} can be read as the set of even numbers. Since there are so many, we can not list them all.
The dots . . . inside the brackets can be read as 'and so on'.

THE EMPTY SET { } or $\emptyset$

The set which has *no members* is called the *empty set.* The set of pupils in your class who are less than 10 centimetres in height is the empty set since there are no members so small.
The empty set is represented by { } or by $\emptyset$.
Here are five examples of the *empty set:*

(1) B is the set of boys each having three noses $B = \emptyset$ or $B = \{\}$
(2) G is the set of girls each of whom is more than 5 metres tall. $G = \emptyset$ or $G = \{\}$
(3) M is the set of months with 46 days. $M = \emptyset$ or $M = \{\}$
(4) L is the set of mice each of which is larger than an elephant. $L = \emptyset$ or $L = \{\}$
(5) C is the set of cows each of which can jump over the moon. $C = \emptyset$ or $C = \{\}$

ASSIGNMENT: Write out another ten *empty sets.*

RING PICTURES

The members of a set can be enclosed by a ring drawing.

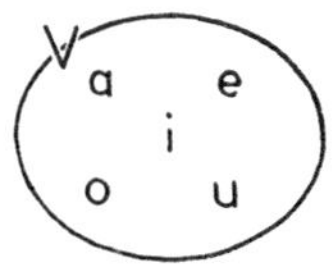

This drawing represents the set V, the set of vowels.
$V = \{a, e, i, o, u\}$

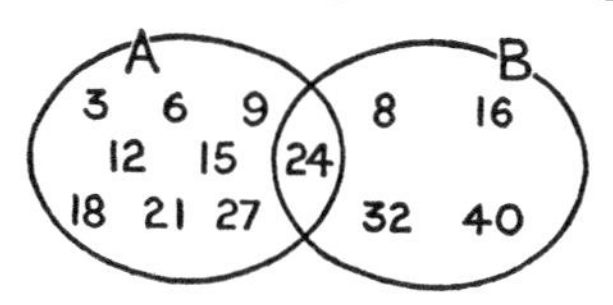

This drawing represents the sets A and B
$A = \{3, 6, 9, 12, 15, 18, 21, 24, 27\}$
$B = \{8, 16, 24, 32, 40\}$
You can see from this drawing that
$24 \in A$ and $24 \in B$

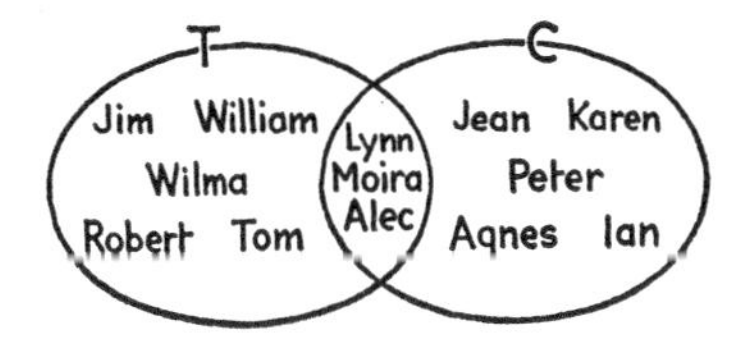

T is the set of pupils in the class each of whom is over 1·50 metres tall.
C is the set of pupils in the class each of whom is good at Arithmetic.

16

From the drawing:

(1) List the set T.
(2) List the set C.
(3) Is Tom over 1·50 metres tall?
(4) Is Tom good at Arithmetic?
(5) Is Jean over 1·50 metres tall?
(6) Is Jean good at Arithmetic?
(7) Is Alec over 1·50 metres tall?
(8) Is Alec good at Arithmetic?
(9) List the set of pupils good at Arithmetic and who are *not* over 1·50 metres tall.
(10) List the set of pupils over 1·50 metres tall and who are *not* good at Arithmetic.
(11) List the set of pupils who are *both* over 1·50 metres tall and are good at Arithmetic.

X is the set of the first ten multiples of 3.
Y is the set of the first seven multiples of 4.

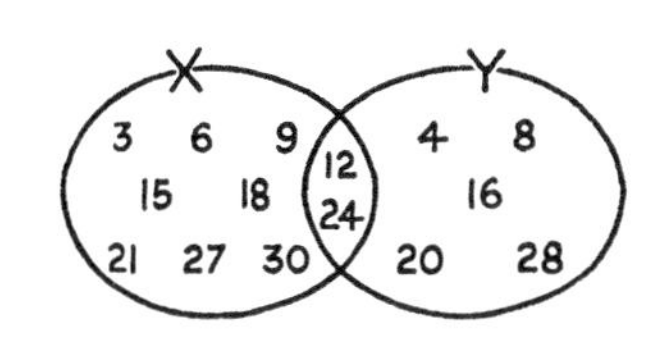

From the drawing:

17

(1) List the set X. (2) List the set Y. (3) Is 20 $\in$ X?
(4) Is 20 $\in$ Y? (5) Is 15 $\in$ X? (6) Is 15 $\in$ Y?
(7) Is 12 $\in$ X? (8) Is 12 $\in$ Y?
(9) List the members of set X which are *not* multiples of 4.
(10) List the members of set Y which are *not* multiples of 3.
(11) List the set of numbers which are members of set X *and also* of set Y.

RING PICTURE PUZZLES

18

(1)

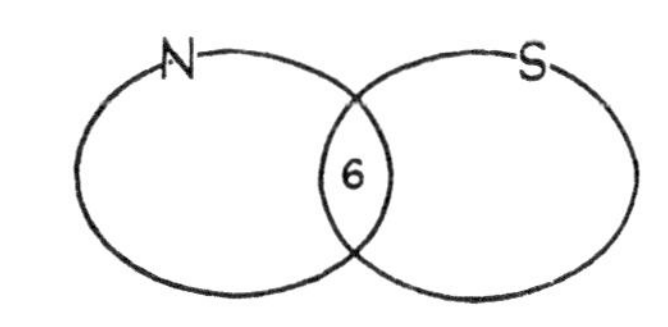

N is the set of 17 girls in the class who like netball.
S is the set of 11 girls in the class who like singing.
6 girls in the class like both netball and singing.

(a) How many girls like netball and do not like singing?
(b) How many girls like singing and do not like netball?
(c) How many girls are involved altogether?

(2) E is the set of boys in Class 7.

F is the set of boys in Class 7 who are members of the School football team.

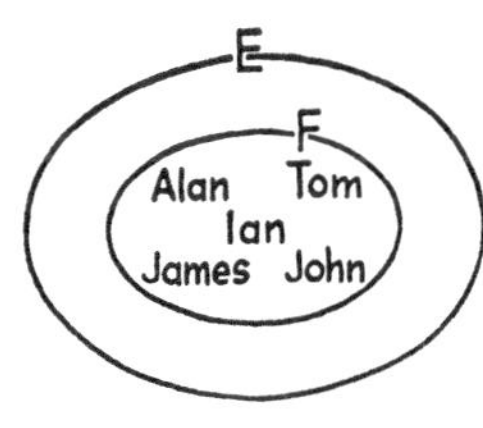

There are 18 boys in Class 7. 5 boys are members of the School football team.

(a) How many boys in Class 7 are not in the School football team?
(b) William is in Class 7. Is William in the School football team?
(c) Andrew is in Class 7. Copy the drawing and write Andrew's name in the correct position.

(3) Draw ring pictures to represent these pairs of sets:

(a) A = {cat, tiger, lion, leopard, dog, bear}
P = {canary, dog, cat, goldfish}
(b) C = {shoes, shirt, trousers, socks, jacket, tie}
S = {shoes, socks}

Multiplication and Division 

Multiplication:

19

(1) 638×2	(2) $9{\cdot}45 \times 5$	(3) 314×7	(4) $86{\cdot}3 \times 9$
(5) 60×43	(6) $5{\cdot}4 \times 80$	(7) 87×64	(8) $3{\cdot}45 \times 37$
(9) 174×52	(10) $2{\cdot}23 \times 28$	(11) 189×76	(12) $18{\cdot}4 \times 19$
(13) 356×67	(14) $4{\cdot}78 \times 45$	(15) 592×89	(16) $6{\cdot}38 \times 34$
(17) 731×34	(18) 864×91	(19) 853×75	(20) 927×86

Division:

20

(1) $537 \div 3$	(2) $36{\cdot}9 \div 9$	(3) $670 \div 5$	(4) $18{\cdot}55 \div 7$
(5) $1\,941 \div 8$	(6) $12{\cdot}4 \div 10$	(7) $4\,282 \div 6$	(8) $78{\cdot}81 \div 10$
(9) $288 \div 36$	(10) $371 \div 53$	(11) $2\,560 \div 40$	(12) $1\,830 \div 30$
(13) $1\,134 \div 21$	(14) $2\,590 \div 35$	(15) $3\,921 \div 64$	(16) $4\,738 \div 75$
(17) $9\,250 \div 37$	(18) $10\,956 \div 83$	(19) $3\,264 \div 72$	(20) $2\,845 \div 67$

PROBLEMS

21

(1) Nine bicycles cost £225. What is the cost of each?

(2) A box holds 76 matches. How many matches will 8 similar boxes hold?

(3) Six chairs cost £50·34. What is the cost of each chair?

(4) A motorist buys 40 litres of petrol each week. How many litres of petrol does he buy in 38 weeks?

(5) 504 cakes were made. The baker packed them 12 to a box. How many boxes did he fill?

(6) When two numbers are multiplied, the result is 3080. If one of these numbers is 70, what is the other?

(7) A truck has 28 sacks on it. Each sack weighs 36·42 kg. What is the load on the truck?

(8) The price of a desk is £4·65. What will 36 such desks cost?

(9) On pay day £775 was shared equally among 25 workmen. How much did each man get?

(10) In 7 weeks a man earns £152·74. How much should he earn in 9 weeks if paid at the same rate?

Speed Test

Complete column (a) first, then column (b), then column (c) and then column (d).

Write answers only. Practice until you can 'beat the clock' – 12 minutes.

22

	(a)	(b)	(c)	(d)
(1)	$3 + 8$	$12 + 6$	$(7 \times 4) + 6$	$5 + (3 \times 9)$
(2)	$5 + 6$	$15 + 4$	$(8 \times 5) + 3$	$(8 \times 7) + 4$
(3)	$7 + 2$	$18 + 7$	$(6 \times 9) + 8$	$(28 \div 7) + 6$
(4)	$4 + 9$	$6 + 29$	$(3 \times 5) - 12$	$8 + (24 \div 6)$
(5)	$9 + 1$	$15 + 9$	$(6 \times 10) - 4$	$14 + (18 \div 3)$
(6)	$8 - 5$	$24 - 8$	$20 - (2 \times 6)$	$(2 \times 8) - 6$
(7)	$12 - 7$	$35 - 6$	$30 - (3 \times 8)$	$(18 \div 2) + 6$
(8)	$14 - 9$	$42 - 3$	$70 - (6 \times 5)$	$(9 \times 4) - 5$
(9)	$16 - 8$	$33 - 9$	$8 + (8 \times 8)$	$(36 \div 6) + 8$
(10)	$13 - 6$	$24 - 11$	$9 + (7 \times 9)$	$(7 \times 6) - 3$
(11)	7×3	7×7	$42 \div 5$	$(4 \times 5) \times 3$
(12)	5×8	5×6	$10 \div 7$	$6 \times (2 \times 5)$
(13)	6×7	7×8	$15 \div 2$	$(12 \div 3) \times 7$
(14)	4×9	9×9	$25 \div 4$	$(10 \times 3) \div 6$
(15)	10×2	6×10	$34 \div 8$	$8 \times (5 \times 2)$
(16)	$35 \div 5$	$63 \div 7$	$29 \div 3$	$(45 \div 5) \times 6$
(17)	$32 \div 8$	$100 \div 10$	$58 \div 9$	$(4 \times 6) \div 8$
(18)	$40 \div 5$	$72 \div 8$	$73 \div 10$	$27 \div (15 \div 5)$
(19)	$48 \div 6$	$45 \div 9$	$35 \div 6$	$40 \div (4 \times 2)$
(20)	$54 \div 9$	$49 \div 7$	$46 \div 7$	$18 \div (24 \div 8)$

ALL THE NINES!

Study the first 10 multiples of 9:
9, 18, 27, 36, 45, 54, 63, 72, 81, 90.
Add the digits of each number. You should get 9.
This provides us with a test as to whether or not a number divides exactly by 9.
The next number which divides exactly by 9 is *99*.
Add the digits. $9 + 9 = 18$. *18* is a multiple of 9 since $1 + 8 = 9$.
Test if *477* is a multiple of 9. $4 + 7 + 7 =$ *18*. 477 *is* a multiple of 9.
Test if *354* is a multiple of 9. $3 + 5 + 4 =$ *12*. 354 *is not* a multiple of 9.

Assignments: (1) Use this test on some large numbers to find if they are divisible by 9 and check your answers by division.
(2) Find tests for discovering multiples of 2, 3, 5 and 10.

B

Decimals—Multiplication

23

Multiplication:

(1) $2{\cdot}9 \times 7$ (2) $29{\cdot} \times 0{\cdot}7$ (3) 46×8 (4) $46 \times 0{\cdot}8$
(5) $64 \times 0{\cdot}9$ (6) $71 \times 0{\cdot}7$ (7) $95 \times 0{\cdot}6$ (8) $67 \times 0{\cdot}5$
(9) $79 \times 0{\cdot}4$ (10) $84 \times 0{\cdot}9$ (11) $63 \times 0{\cdot}6$ (12) $83 \times 0{\cdot}8$
(13) 46×3 (14) $46 \times 0{\cdot}4$ (15) $(46 \times 3) + (46 \times 0{\cdot}4)$
(16) 71×8 (17) $71 \times 0{\cdot}5$ (18) $(71 \times 8) + (71 \times 0{\cdot}5)$

Find the answers to:

(19) 9 times 86 (20) $0{\cdot}3$ times 86
(21) $(86 \times 9) + (86 \times 0{\cdot}3)$ (22) 46 times $3{\cdot}4$
(23) 71 times $8{\cdot}5$ (24) $86 \times 9{\cdot}3$
(25) $(43 \times 5) + (43 \times 0{\cdot}9)$ (26) $(75 \times 8) + (75 \times 0{\cdot}6)$
(27) $(62 \times 4) + (62 \times 0{\cdot}8)$ (28) $(58 \times 7) + (58 \times 0{\cdot}5)$

Write down the number in the box:

(29) $43 \times 5{\cdot}9 = \square$ (30) $75 \times 8{\cdot}6 = \square$
(31) $62 \times 4{\cdot}8 = \square$ (32) $58 \times 7{\cdot}5 = \square$

ABE TRIES ANOTHER METHOD

4·5 × 34

4·5
× 34·
135· ×30
18·0 × 4
153·0

Line up points

34·
×4·5
×4 136·
×0·5 17·0
153·0

34 × 4·5

24

Multiplication:

(1) $36 \times 2{\cdot}3$ (2) $2{\cdot}3 \times 36$ (3) $41 \times 4{\cdot}6$
(4) $61 \times 5{\cdot}2$ (5) $73 \times 6{\cdot}3$ (6) $54 \times 7{\cdot}5$
(7) $88 \times 7{\cdot}4$ (8) $63 \times 8{\cdot}4$ (9) $78 \times 7{\cdot}6$
(10) $94 \times 6{\cdot}2$ (11) $22 \times 5{\cdot}3$ (12) $49 \times 7{\cdot}4$
(13) $52 \times 4{\cdot}3$ (14) $60 \times 5{\cdot}2$ (15) $80 \times 7{\cdot}3$
(16) $73 \times 5{\cdot}8$ (17) $88 \times 6{\cdot}6$ (18) $82 \times 8{\cdot}7$
(19) $65 \times 7{\cdot}8$ (20) $74 \times 8{\cdot}9$

ABE TRIES HARDER

4·2 × 142

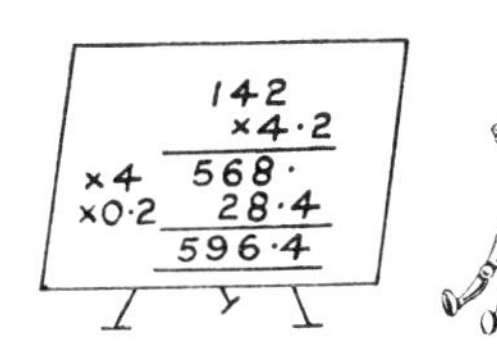

142 × 4·2

Multiplication:

25

(1) 132 × 4·2	(2) 212 × 3·5	(3) 312 × 2·3
(4) 412 × 3·2	(5) 526 × 4·3	(6) 632 × 5·3
(7) 326 × 3·8	(8) 625 × 7·1	(9) 826 × 8·3
(10) 204 × 6·2	(11) 340 × 7·3	(12) 804 × 5·4

PROBLEMS

26

(1) A dress costs £7·20. How much will 34 dresses cost altogether, if they all cost the same?

(2) How much will 63 bags of potatoes weigh if each bag contains 4·2 kg. of potatoes?

(3) A bag of sugar weighs 0·5 kg. What is the total weight of 39 such bags of sugar?

(4) A box weighs 3·6 kg. How much will 96 similar boxes weigh altogether?

(5) A strip of wood measures 2·8 metres. What will be the total length of 57 such strips?

ABE GETS TOUGH

23·0 × 15·2

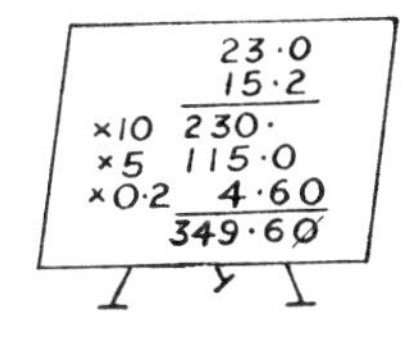

Multiplication:

27

(1) 38·0 × 13·3	(2) 29·0 × 15·6	(3) 55·0 × 14·2
(4) 64·0 × 15·3	(5) 77·0 × 21·3	(6) 85 × 22·4
(7) 92·0 × 24·5	(8) 68 × 37·1	(9) 46 × 53·2
(10) 49 × 19·2	(11) 76 × 26·8	(12) 87 × 61·2
(13) 48·0 × 16·2	(14) 72 × 46·2	(15) 54 × 65·7
(16) 69 × 59·4	(17) 81 × 92·8	(18) 67 × 85·7

ABE HAS PROBLEMS

28

(1) Twenty-five barrels each weigh 36·3 kg. What is their total weight?

(2) 36 people saved £12·80 each. How much did they save altogether?

(3) $(63 \times 2{\cdot}3) + (36 \times 12{\cdot}4) = \square$. Find $\square$.

(4) How much milk is there in 36 full bottles, if each bottle holds 2·2 litres.

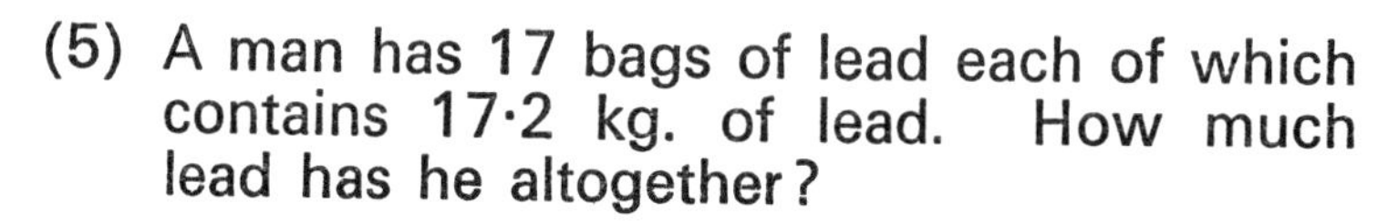

(5) A man has 17 bags of lead each of which contains 17·2 kg. of lead. How much lead has he altogether?

ABE'S PUZZLERS

Copy and fill in the missing figures denoted by *.

29

(1)
```
   26·
  ×2·4
  ----
   5*·
  10·4
  ----
  6*·*
```

(2)
```
   38·
  ×7·2
  ----
  2*6·
    *·6
  ----
  ***·6
```

(3)
```
   3*·
  ×6·3
  ----
  198·
    9·*
  ----
  ***·*
```

(4)
```
   4*·
  ×0·2
  ----
   *·4
```

(5)
```
   7*·
  ×0·8
  ----
  **·6
```

(6)
```
   **·
   5·3
  ----
  265·
   **·*
  ----
  ***·*
```

Find the replacement for $\square$ in each example:

(7) $46{\cdot}2 \times 6 = \square$

(8) $24{\cdot}3 \times 8 = \square$

(9) $23 \times 0{\cdot}4 = \square$

(10) $36 \times 0{\cdot}7 = \square$

(11) $52 \times 0{\cdot}3 = \square$

(12) $71 \times 0{\cdot}5 = \square$

(13) $29 \times 0{\cdot}8 = \square$

(14) $51 \times 0{\cdot}9 = \square$

(15) $29 \times 4 = \square$

(16) $51 \times 7 = \square$

(17) $29 \times 4{\cdot}8 = \square$

(18) $51 \times 7{\cdot}9 = \square$

ABE HAS A PROBLEM – MULTIPLICATION

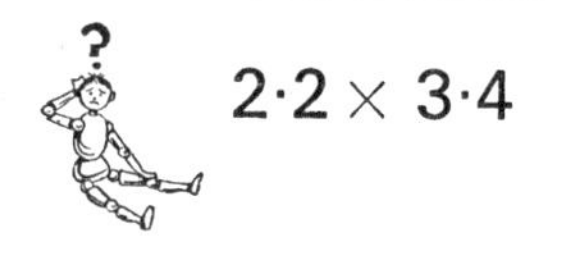

$2{\cdot}2 \times 3{\cdot}4$

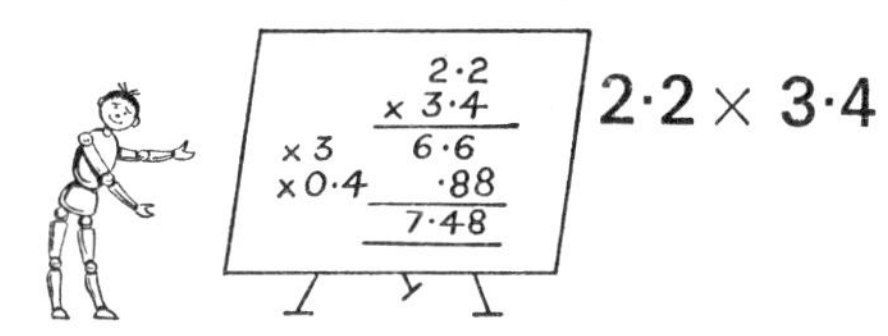

$2{\cdot}2 \times 3{\cdot}4$

Multiplication:

30

(1) $1{\cdot}3 \times 2$ (2) $1{\cdot}3 \times 0{\cdot}3$ (3) $1{\cdot}3 \times 2{\cdot}3$
(4) $2{\cdot}4 \times 4$ (5) $2{\cdot}4 \times 0{\cdot}2$ (6) $2{\cdot}4 \times 4{\cdot}2$
(7) $5{\cdot}4 \times 3{\cdot}4$ (8) $7{\cdot}2 \times 4{\cdot}3$ (9) $68 \times 5{\cdot}2$

Find the answers to:
(10) $(6{\cdot}4 \times 5) + (6{\cdot}4 \times 0{\cdot}7)$ (11) $6{\cdot}4 \times 5{\cdot}7 = \square$
(12) $(8{\cdot}8 \times 4) + (8{\cdot}8 \times 0{\cdot}2)$ (13) $8{\cdot}8 \times 4{\cdot}2 = \square$

State which of these statements are true:

31

(1) $4 \times 2 = 8$ (2) $4 \times 0{\cdot}2 = 0{\cdot}8$
(3) $0{\cdot}4 \times 2 = 0{\cdot}08$ (4) $2{\cdot}2 \times 6 = 12{\cdot}2$
(5) $3{\cdot}4 \times 0{\cdot}2 = 0{\cdot}68$ (6) $4{\cdot}2 \times 8 = 32{\cdot}16$
(7) $3{\cdot}2 \times 2{\cdot}2 = 7{\cdot}04$ (8) $4{\cdot}1 \times 2{\cdot}3 = 8{\cdot}43$
(9) $5{\cdot}8 \times 0{\cdot}8 = 0{\cdot}464$ (10) $6{\cdot}2 \times 0{\cdot}4 = 1{\cdot}28$
(11) $7{\cdot}3 \times 0{\cdot}5 = 36{\cdot}5$ (12) $7{\cdot}4 \times 0{\cdot}6 = 44{\cdot}4$

More Multiplication and Brackets:

32

(1) $7{\cdot}6 \times 8{\cdot}7$ (2) $5{\cdot}2 \times 2{\cdot}3$ (3) $1{\cdot}3 \times 4{\cdot}2$ (4) $2{\cdot}2 \times 3{\cdot}4$
(5) $(7{\cdot}8 \times 8{\cdot}7) + (5{\cdot}2 + 2{\cdot}3)$ (6) $(4{\cdot}2 \times 1{\cdot}3) + (3{\cdot}4 \times 2{\cdot}2)$
(7) $(4{\cdot}6 \times 5{\cdot}3) + (6{\cdot}7 \times 1{\cdot}2)$ (8) $6{\cdot}8(8{\cdot}4 + 7{\cdot}6)$
(9) $6{\cdot}2(0{\cdot}6 + 0{\cdot}4)$ (10) $14{\cdot}6 \times 3{\cdot}2$
(11) $21{\cdot}3 \times 4{\cdot}6$ (12) $32{\cdot}2 \times 5{\cdot}6$ (13) $23{\cdot}8 \times 6{\cdot}7$
(14) $2{\cdot}6(3{\cdot}4 + 6{\cdot}6) + 10(3{\cdot}3 + 5{\cdot}6) + 24$

ABE HAS PROBLEMS

33

(1) A motorist puts 24 litres of petrol in his car. If petrol costs 7·5p/litre, how much change will he get from £2?
(2) A train travels for 2·5 hours at 45 km/hour.
How many kilometres has the train travelled?

(3) A colour transparency is 2·5 cm long and 1·9 cm broad. What is the area of the transparency?
(4) If one litre of gas weighs 0·3g what will be the weight of 22·5 litres?
(5) A householder uses 114 units of electricity. If he pays 2·5p/unit for the first 36 units and 1·25p/unit for the remainder, calculate the total bill.

Abe and his Accounts

Copy and complete these accounts:

34

(1) From : John Jones – Grocer, 9 Main St.
To : Abe, 0 Nowhere St., Somewhere

		£
2·5 kg Potatoes at 4p per kg	=	
1·5 kg Sugar at 7p per kg	=	
0·5 kg Tea at 49p per kg	=	
2·5 litre Milk at 10p per litre	=	
Total	=	

(2) From : Joseph Beef – Butcher, 6 Ayr St.
To : Abe, 0 Nowhere St., Somewhere

		£
1·75 kg Sausages at 16p per kg	=	
0·75 kg Frying Steak at 55p per $\frac{1}{2}$ kg	=	
0·25 kg Bacon at 72p per kg	=	
1·25 kg Pork at 80p per kg	=	
Total	=	

(3) From : John Wood – Hardware, 6 Timber Yard
To : Abe, 0 Nowhere St., Somewhere

		£
3·2m wood at $12\frac{1}{2}$p/metre	=	
4m formica at $21\frac{1}{2}$p/metre	=	
2 hammers at $52\frac{1}{2}$p each	=	
3·2 litres varnish $42\frac{1}{2}$ p/litre	=	
Total	=	

(4) From : H. O. Tell, 6 Spring St.
To : Abe, 0 Nowhere St., Somewhere

		£
10 Lunches at $47\frac{1}{2}$p each	=	
15 Dinners at $52\frac{1}{2}$p each	=	
9 Breakfasts at $25\frac{1}{2}$p each	=	
36 Coffees at $6\frac{1}{2}$p each	=	
Total	=	

(5) Post Office

Post Office

		£
24 stamps at $2\frac{1}{2}$p each	=	
12 stamps at $12\frac{1}{2}$p each	=	
16 Insurance Stamps at £1·15 each	=	
Total	=	

(6) From : T. Aire – The Garage, Blower St.
To : Abe, 0 Nowhere St., Somewhere

		£
90 litres Oil at $14\frac{1}{2}$p/litre	=	
75·5 litres Petrol at 32p/litre	=	
6 Tyres at £8·75 each	=	
Total	=	

(7) From : Angela's, 8 Tone St.
To : Mrs. Abe, 0 Nowhere St., Somewhere

		£
7 Dresses at £5·25 each	=	
14 pairs Nylon tights at $42\frac{1}{2}$p pair	=	
6 pairs Slacks at £4·35 each pair	=	
Total	=	

(8) From : A. Boot, 7 Shorn St.
To : Abe, 0 Nowhere St., Somewhere

		£
12 pairs shoes at £4·25/pair	=	
8 pairs sandals at £1·36/pair	=	
15 pairs slippers at £0·75/pair	=	
Total	=	

MORE ABE PROBLEMS

35

(1) A football team played 32 games and had an average of 2·5 goals for and 0·5 goals against. How many goals did they score and how many did they lose?

Games	W	L	D	For	Ag'st
32	29	2	1	?	?

(2) How much does a man earn in a 40 hour week if he is paid 42·5p/hour?

(3) 16·5 m² of carpet are required to cover the floor of a room. Calculate the cost at £1·65/m².

(4) 

What is the weekly wage bill for a firm employing 38 men who earn £16·25 each per week?

(5) A car travels 16·25 km on one litre of petrol. How far will it travel on 8 litres of petrol?

Number Patterns

NATURAL NUMBERS

The set of natural numbers N can be written thus:

$N = \{1, 2, 3, 4, 5, 6, 7, 8, 9, 10, 11, \ldots\}$

Each natural number can be represented by a *dot pattern*.

The first 6 natural numbers are represented on a die thus:

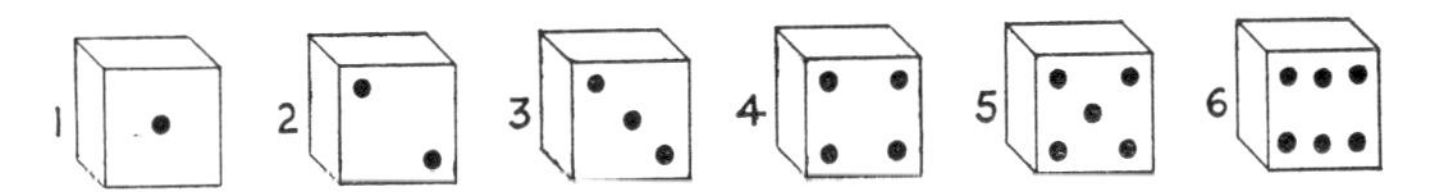

Make your own dot patterns to represent:

36

(1) 8 (2) 9 (3) 10 (4) 15 (5) 7 (6) 12
(7) 14 (8) 11 (9) 16 (10) 18 (11) 25 (12) 20

RECTANGULAR NUMBERS

If a number can be represented by equal rows of dots, a rectangle may be formed by joining up the dots as shown. Such a number is called a *Rectangular Number*.

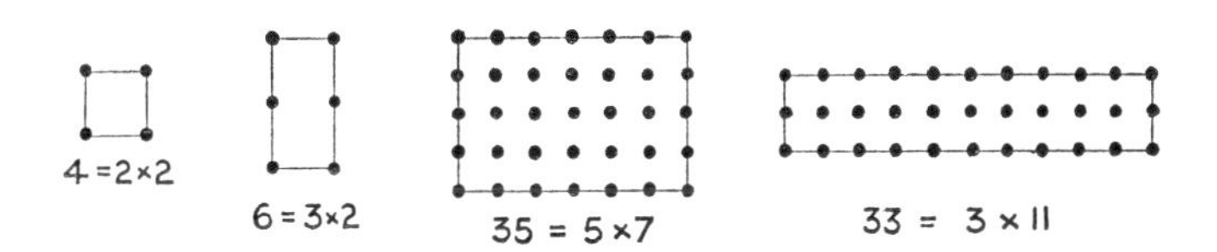

The numbers 4, 6, 35 and 33 are rectangular numbers

PRIME NUMBERS

Any number which can not be represented by an equal-rows rectangular dot pattern is called a *Prime Number*. 3 and 23 are prime numbers.

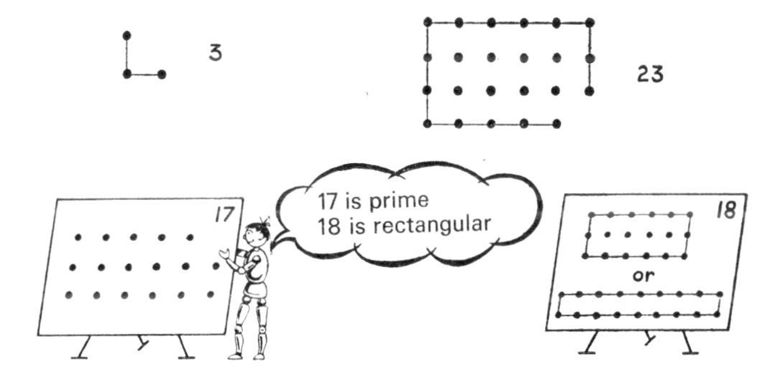

37

(1) Draw 'equal rows' rectangular dot patterns to show that each of these numbers is rectangular:
(a) 8 (b) 9 (c) 10 (d) 12 (e) 14 (f) 15
(g) 16 (h) 20 (i) 21 (j) 22 (k) 24

(2) Copy and complete the set of rectangular numbers less than 36.
R = {4, 6, 8, 9, 10, 12, 14, 15, 16, 18, 20, . . .}

(3) Draw dot patterns to represent each of these prime numbers:
(a) 2 (b) 5 (c) 7 (d) 11 (e) 13 (f) 19

(4) Copy and complete the set of prime numbers less than 36.
P = {2, 3, 5, 7, 11, 13, 17, 19, , , 31}

(5) There is a natural number which is neither prime nor rectangular. Find it.

FACTORS

Natural numbers can be drawn in more than one way. Thus 10 may be drawn as follows:

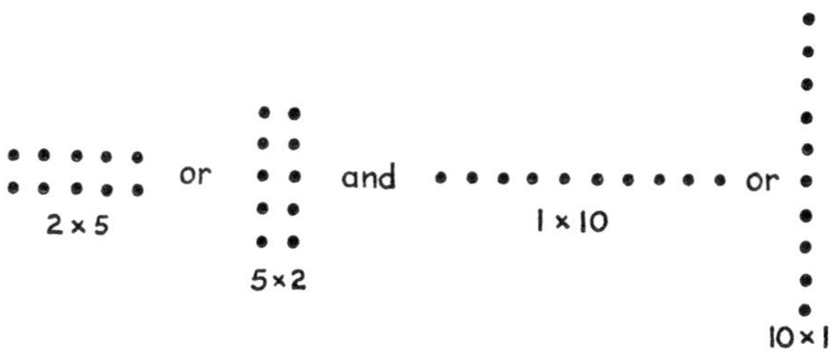

Possible pairs of factors of 10 are *2 and 5* and *1 and 10*.

The set of factors of 10 is F = {1, 2, 5, 10}

38

(1) Draw dot patterns to represent each number and find the set of factors as on Abe's board:
(a) 6 (b) 4 (c) 9 (d) 3 (e) 5 (f) 14
(g) 2 (h) 8 (i) 11 (j) 7 (k) 18 (l) 16

(2) The numbers 2, 3, 5, 7 and 11 are either prime or rectangular. Which?

(3) How many members has the set of factors in each of 2, 3, 5, 7 and 11?

(4) One factor is the number itself in each of 2, 3, 5, 7 and 11. What is the other factor?

(5) Copy and complete: A prime number has only — factors which are — — — — —.

(6) The numbers 4, 6, 8, 9, 10, 12, 14, 16 and 18 are either prime or rectangular. Which?
(7) How many factors have the numbers 4 and 9?
(8) How many factors have the numbers 6, 8, 10, 14 and 18?
(9) Copy and complete: A rectangular number has more than — —.

SQUARE NUMBERS

Look at these dot patterns:

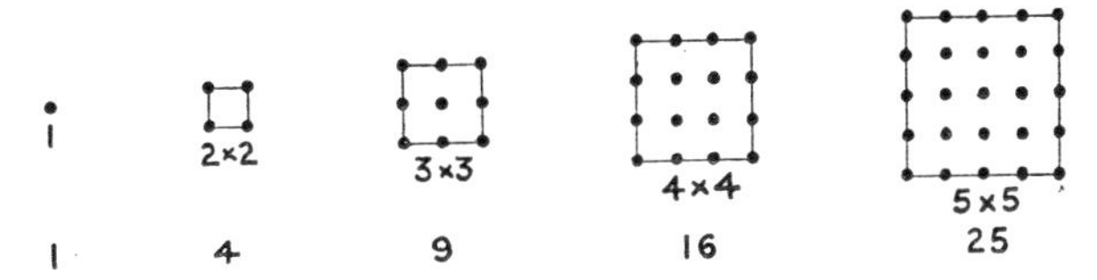

{1, 4, 9, 16, 25} is the set of the *first five square numbers*.
Square numbers have the same number of rows as columns.

NOTE: 5×5 may be written 5^2.
5^2 is read as 'five *squared*'.
$5^2 = 25$, $4^2 = 16$, $3^2 = 9$ and $2^2 = 4$.

39

(1) Find the value of:
(a) 6^2 (b) 7^2 (c) 8^2 (d) 9^2 (e) 10^2 (f) $3^2 + 4^2$ (g) $6^2 + 8^2$
(2) Represent the first five square numbers as dot patterns.
(3) Draw dot patterns to represent the next five square numbers.
(4) Find a square number which is not a rectangular number.

TRIANGULAR NUMBERS

Look at these dot patterns:

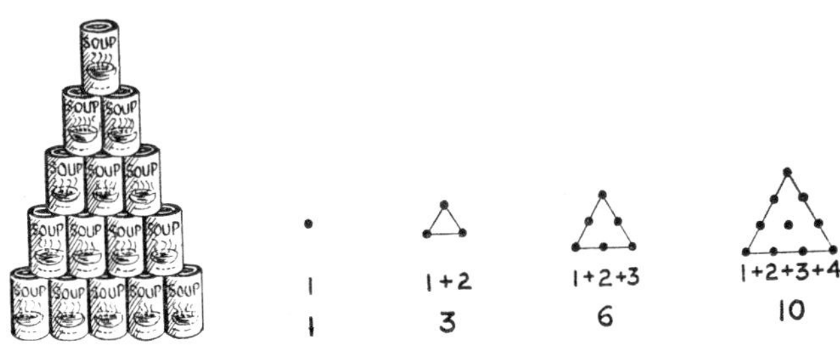

40

{1, 3, 6, 10, 15} is the set of the *first five triangular numbers.*

(1) Represent the first four triangular numbers as dot patterns.
(2) How many dots are there in the bottom row of each of these triangular numbers: (a) 3? (b) 6? (c) 10? (d) 15?
(3) Draw the fifth triangular number in dot form.
(4) Draw the next five triangular numbers in dot form.
(5) Copy and complete: 1, 3, 6, 10, 15, —, —, —, —, 55.
(6) Find a prime triangular number.
(7) Find a triangular number which is neither prime nor rectangular.

Missing Numbers

Study the pattern of the numbers in this table:

0	1	2	3	4
5	6	7	8	9
10	11	12	13	14
15	16	17	18	19
20	21	22	23	24

41 (1) Copy the 24 table on to squared paper.
Abe has drawn a straight line through the *first 4 multiples of 6.*
(2) Draw a straight line through the first four multiples of 5.
(3) Draw a straight line through the first five multiples of 4.

From your table, copy and complete the sequences:
(4) 0, 6, 12, 18, — (5) 0, 5, 10, —, — (6) 4, 8, 12, —, —
(7) 1, 7, 13, — (8) 5, 6, —, —, 9 (9) 3, 8, 13, —, —

Copy and complete the sequences:
(10) 1, 2, 3, 4, —, — (11) 2, 4, 6, —, —, 12
(12) 1, 3, 5, 7, —, — (13) 7, 14, 21, 28, —, —
(14) 1^2, 2^2, 3^2. —, —, — (15) 1, 4, 9, 16, —, —
(16) 10, 20, 30, —, —, 60 (17) 5, 10, —, 20, —, —
(18) 1, 3, 6, 10, —, — (19) 1, 2, 4, 8, —, —
(20) 1, 10, 100, 1000, —, —

MAGIC SQUARES

42 Copy and complete the magic squares:

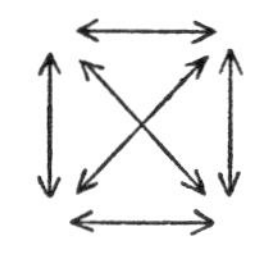

(1)

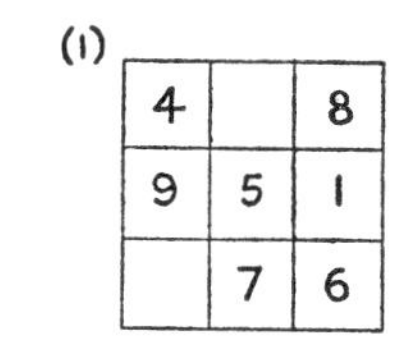

4		8
9	5	1
	7	6

(2)

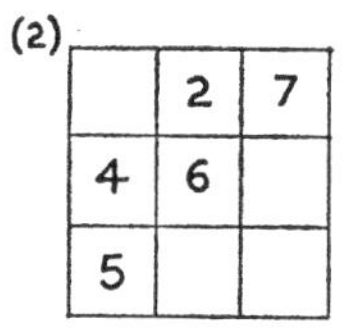

	2	7
4	6	
5		

(3)

8	9	
	7	11
		6

(4)

4	9		16
14	7		2
15	6		3
1	12		13

(5)

4	15	6	
		3	16
11		13	2
		12	7

(6)

	16	22	3	9
8		20	21	2
1	7	13		
24		6	12	18
	23	4	10	

Progress Checks

PROGRESS CHECK 1

Find the answers to:

(1) Write down the set of the first six multiples of 7.

(2) Write down the set of prime numbers between 30 and 40.

(3) Copy and complete: 1, 3, 6, 10, —, —

(4) 2 345 + 1 189 + 4 783

(5) 83 362 − 17 835

(6) 37 × 42

(7) 3 536 ÷ 52

(8) 47 × 0·4

(9) 62 × 12·5

(10) 3·6 × 2·7

PROGRESS CHECK 2

Find the answers to:

(1) Write down the set of factors of 32.

(2) Write down the set of rectangular numbers between 30 and 40.

(3) Copy and complete: 1, 4, 9, 16, —, —

(4) 14·6 + 72·8 + 33·9

(5) 10,000 − 356

(6) 56 × 39

(7) 2 952 ÷ 41

(8) 78 × 0·9

(9) 56 × 23·4

(10) 4·8 × 3·3

PROGRESS CHECK 3

(1) What number is ten less than one million?

(2) A digital clock reads | 16 24 | What is the a.m. or p.m. time?

(3) A van weighs 1 825 kg when empty and 3 005 kg when fully loaded. Find the weight of its load.

(4) The price of a record-player is £24·50. Find the cost of 9 such record-players.

(5) If 18 similar boxes hold 828 matches, how many matches does each box hold?

(6) How much will 25 bags of potatoes weigh if each bag weighs 3·4 kg?

(7) An aeroplane travels for 2·5 hours at 680 km/hour. How far does the plane travel?

(8) Find the total cost of 8 Breakfasts at 28·5p each, 6 Lunches at 52·5p each and 8 Teas at 14p each.

(9) What is the smallest prime number greater than 100?

(10) Write in words: 3 010 600.

Estimations—Rounding off Numbers

ASSIGNMENTS:

(1) A "metre stick" is 100 centimetres long. Think of this and estimate your teacher's height.

(2) Estimate (do not measure) in centimetres the length and breadth of this page of your book. Write your answers and check with your ruler.

(3) Estimate in metres the length and breadth of your classroom. Check by measurement.

(4) Estimate in metres the height of your classroom.

(5) Estimate in suitable units, and check where possible:

- (a) The length of your step.
- (b) The weights in grammes or kilogrammes of various objects and pupils in the classroom.
- (c) The liquid measure in litres of any containers in the classroom.

Approximate answers may be obtained by *ROUNDING OFF* the numbers involved. For example if an approximate total attendance at 4 football matches is required:

	exact		approx.
add	3 125	$\doteqdot$	3 000
	7 928	$\doteqdot$	8 000
	5 207	$\doteqdot$	5 000
	9 856	$\doteqdot$	10 000
	Total	$\doteqdot$	26 000

where $\doteqdot$ means "is approximately equal to"

	exact		approx.
also	385	$\doteqdot$	400
	$\times 18$	$\doteqdot$	$\times 20$
	answer	$\doteqdot$	8 000

and $\frac{1}{2}$ of £7·95 $\doteqdot$ £4·00

Study these examples and the explanations which follow.

ROUNDING OFF

In the above examples,

£7·95 is nearer £8·00 than it is to £7·00.

£7·95 rounded off to the nearest £ is £8·00.

We write £7·95 $\doteqdot$ £8·00 to the nearest £.

Similarly, 18 $\doteqdot$ 20 to the nearest 10.

385 $\doteqdot$ 400 to the nearest 100.

7 928 $\doteqdot$ 8 000 to the nearest 1 000.

3 125 $\doteqdot$ 3 000 to the nearest 1 000.

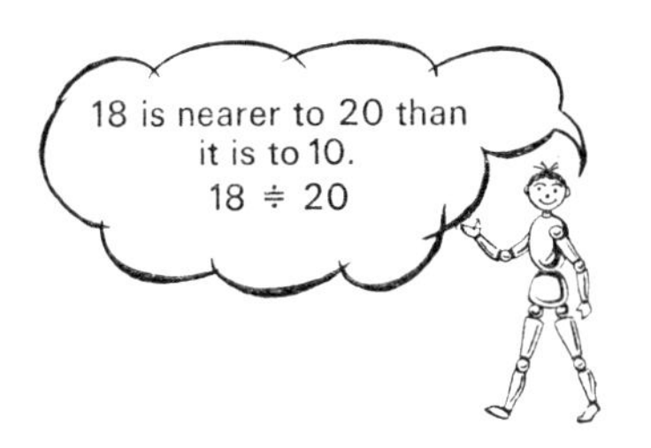

Note: If the number is exactly mid-way between the limits, then round *UP:*

3·5 $\doteqdot$ 4 to the nearest unit.

65 $\doteqdot$ 70 to the nearest 10.

750 $\doteqdot$ 800 to the nearest 100.

43

Round off to the nearest unit:

(1) 6·9 (2) 3·7 (3) 5·2 (4) 10·4 (5) 13·8 (6) 14·25

Round off to the nearest ten:

(7) 37 (8) 53 (9) 99 (10) 137 (11) 653 (12) 299
(13) 203 (14) 106 (15) 111 (16) 86 (17) 94 (18) 79

Round off to the nearest hundred:

(19) 137 (20) 653 (21) 299 (22) 203 (23) 106 (24) 792
(25) 1 563 (26) 1 635 (27) 4 106
(28) 4 160 (29) 7 984 (30) 9 990

Round off to the nearest thousand:

(31) 5 800 (32) 6 200 (33) 3 194 (34) 1 914
(35) 9 501 (36) 10 105 (37) 19 919 (38) 33 099
(39) 90 499 (40) 99 700

(41) Round off 5 555·5 to the nearest (a) unit, (b) ten, (c) hundred, (d) thousand.

ESTIMATION—APPROXIMATION

Round off to the nearest £:

44

(1) £13·75 (2) £8·40 (3) £9·55 (4) £17·50 (5) £2·05
(6) £19·95 (7) £1·45 (8) £20·20 (9) £3·10 (10) £0·60

Round off to the nearest centimetre:

(11) 7·3 cm (12) 4·9 cm (13) 12·5 cm (14) 25·6 cm
(15) 29·8 cm

Round off to the nearest metre:

(16) 5·80 m (17) 8·36 m (18) 49·75 m (19) 35·50 m
(20) 27·05 m

Round off to the nearest kilometre:

(21) 2·900 km (22) 9. 550 km (23) 7·495 km (24) 13·095 km
(25) 10·5 km

PROBLEMS

45

(1) Round off to the nearest 1 000 each of the numbers, 5 425, 6 982, 7 407 and 9 856. Add the numbers you get.

(2) Add the numbers, 5 425, 6 982, 7 407, 9 856. Round off your *answer* to the nearest 1 000.

(3) An approximate answer is required for 86×24. Round off each number to the nearest 10 and find this answer.

(4) An approximate answer is required for 523×76. Round off the numbers suitably and find this answer.

(5) Multiply 7·9 by 3·14. Find an approximate answer by first rounding off each number to the nearest unit.

(6) Find $\frac{1}{3}$ of £122·55. Round off the answer to the nearest £.

(7) By rounding off both numbers to the nearest unit find an approximate answer for $17{\cdot}8 \div 2{\cdot}7$.

More Multiplication and Decimal Approximation

ABE TRIES MORE DECIMALS – MULTIPLICATION

4·25 × 3·25

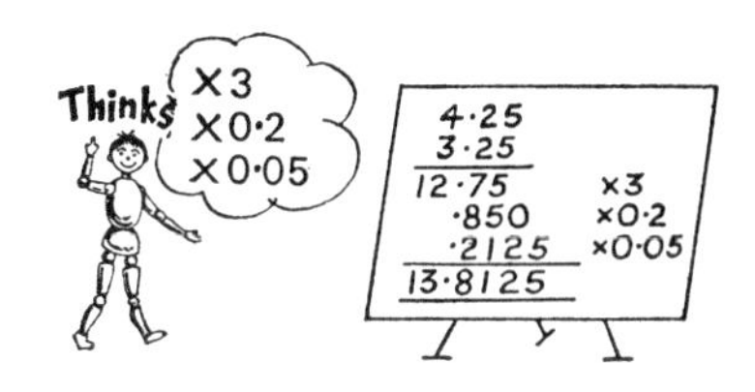

Multiplication:

(1) 2·12 × 3·23 (2) 4·23 × 3·25 (3) 5·16 × 2·34
(4) 6·32 × 4·14 (5) 6·42 × 5·32 (6) 7·18 × 6·28
(7) 8·25 × 6·82 (8) 5·54 × 9·27 (9) 9·16 × 7·83
(10) 7·68 × 8·34

46

Abe Thinks Things Out – Approximation

4·25 m × 3·25 m = 13·8125 m²

Abe Decides

He says read

8·2356 as 8·236 8·2352 as 8·235

This is called approximation to the 3rd decimal place.

Note: If the fourth decimal place is 5 or more add 1 to third decimal place.

So 13·8125m² is 13·813m² approximately.

c

Abe answers questions:

Give an approximate for:

47 (1) 14·168 to the second decimal place

(2) 26·23672 to the third decimal place

(3) 4·281 to the second decimal place

(4) 7·3265 to the first decimal place

APPROXIMATION – DECIMALS

Give an approximation:

48
(1) To the Third decimal place for 2·8168
(2) To the Second decimal place for 4·824
(3) To the First decimal place for 3·281
(4) To the Second decimal place for 19·687
(5) To the Third decimal place for 24·8125
(6) To the First decimal place for 68·872
(7) To the Second decimal place for 4·635
(8) To the Second decimal place for 17·838
(9) To the Third decimal place for 3·8927
(10) To the First decimal place for 7·38

Abe Thinks **Hard!**

36·0 is correct to the "tenths" place or the first decimal place.
36 is correct to the "units" place.
36·00 is correct to the "hundredths" place or the second decimal place.

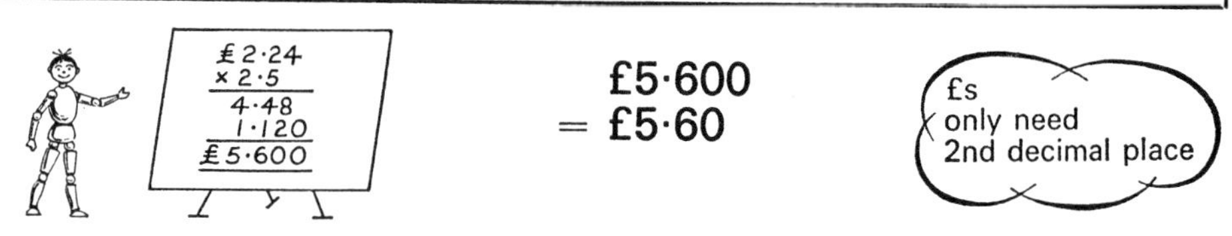

Remember: Smallest Weight Unit is 1 gramme = 0·001 kg
Smallest Money Unit is 1 new penny = £0·01
Smallest Length Unit is 1 millimetre = 0·001 metre

49

(1) Express £13·5612 to the nearest penny.

(2) Express 3·216 m to the nearest centimetre (2nd decimal place).

(3) Express 41·2356 kg to the nearest gramme (3rd decimal place).

(4) Express £2·35 × 2 to the nearest penny.

(5) Multiply 4·625 m by 4 and express the answer to the third decimal place.

ROUNDING OFF DECIMALS

50

(1) Round off to the first decimal place:

(a) 3·69 (b) 3·61 (c) 3·65 (d) 3·66 (e) 7·42
(f) 9·76 (g) 2·50 (h) 10·09 (i) 23·23 (j) 8·96
(k) 7·86 (l) 9·97

(2) Round off each answer to the first decimal place:

(a) 2·35 + 3·94 + 4·8 + 9·07 + 7·3 (b) 73·4 − 29·55
(c) 8·7 × 3·1 (d) 13·70 ÷ 5 (e) $\frac{1}{3}$ of 103·65

(3) In four tests Abe scored 36, 39, 34 and 34. Find his total of marks. Divide the total of marks by 4 and give this answer correct to the first decimal place.

(4) A piece of ground is 14·2 metres long and 8·3 metres broad. Find its area in square metres. Write your answer to the first decimal place.

(5) A car travels 380 km and uses 32 litres of petrol. How far does the car travel on one litre of petrol? Write your answer to the first decimal place.

H.C.F. and L.C.M.

HIGHEST COMMON FACTORS

HCF

Study the following factors for 12 and for 18.

$12 = 12 \times 1 = 6 \times 2 = 4 \times 3.$ $F = \{1, 2, 3, 4, 6, 12\}$
$18 = 18 \times 1 = 9 \times 2 = 6 \times 3.$ $G = \{1, 2, 3, 6, 9, 18\}$

The numbers 12 and 18 share the factors 1, 2, 3 and 6.

The common factors of 12 and 18 are 1, 2, 3, and 6

We say, "The Highest Common Factor of 12 and 18 is 6".
We write, "HCF of 12 and 18 is 6".

Abe finds the HCF of 40 and 100.

Use the method on Abe's board to find the HCF of each pair of numbers:

51

(1) 12 and 16 (2) 16 and 24 (3) 18 and 24 (4) 20 and 50
(5) 15 and 100 (6) 18 and 36 (7) 60 and 100 (8) 35 and 49
(9) 18 and 20 (10) 25 and 50

Write down the HCF of each pair of numbers with or without working:

(11) 8 and 10 (12) 12 and 20 (13) 10 and 50
(14) 25 and 35 (15) 40 and 56 (16) 27 and 63
(17) 20 and 100 (18) 25 and 100 (19) 50 and 100
(20) 24 and 36

USING HCFs IN FRACTIONS

$\frac{75}{100}$

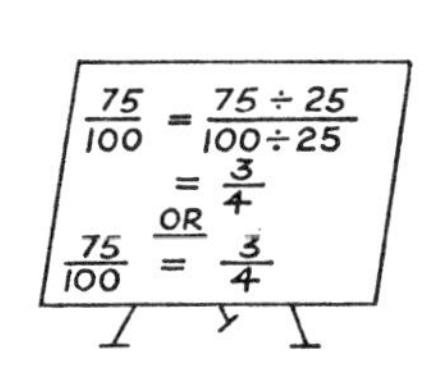

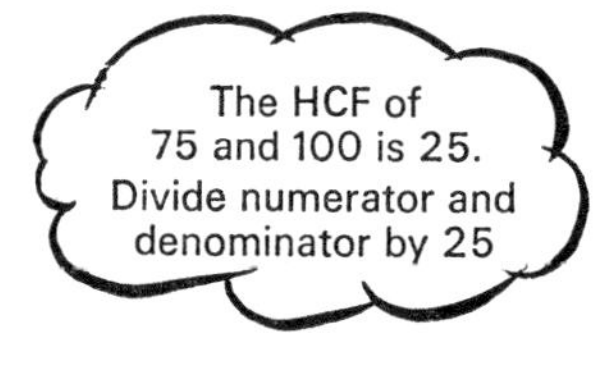

Express these fractions in their lowest terms by dividing numerator and denominator by the HCF:

52

(1) $\frac{3}{9}$ (2) $\frac{5}{20}$ (3) $\frac{7}{42}$ (4) $\frac{6}{60}$ (5) $\frac{8}{12}$
(6) $\frac{10}{25}$ (7) $\frac{6}{27}$ (8) $\frac{24}{42}$ (9) $\frac{27}{45}$ (10) $\frac{24}{32}$
(11) $\frac{70}{100}$ (12) $\frac{27}{36}$ (13) $\frac{28}{63}$ (14) $\frac{14}{22}$ (15) $\frac{5}{100}$
(16) $\frac{20}{100}$ (17) $\frac{50}{100}$ (18) $\frac{35}{100}$ (19) $\frac{25}{100}$ (20) $\frac{60}{100}$

LEAST COMMON MULTIPLES

Study the multiples of 4 and 6 which are less than 50.

4: M = {4, 8, *12*, 16, 20, *24*, 28, 32, *36*, 40, 44, *48*}

6: **N** = {6, *12*, 18, *24*, 30, *36*, 42, *48*}

The numbers 4 and 6 have in common the multiples 12, 24, 36 and 48.
The common multiples of 4 and 6 are 12, 24, 36 and 48.
We say, "The Least Common Multiple of 4 and 6 is 12".
We write, "LCM of 4 and 6 is 12".

Abe finds the LCM of 6 and 15.

Use the method on Abe's board to find the LCM of each pair of numbers:

53

(1) 2 and 3 (2) 3 and 5 (3) 4 and 10 (4) 6 and 8
(5) 8 and 10 (6) 3 and 8 (7) 9 and 15 (8) 20 and 30
(9) 8 and 12 (10) 20 and 25

Write down the LCM of each pair of numbers with or without working:

(11) 6 and 9 (12) 2 and 7 (13) 4 and 5 (14) 6 and 10
(15) 3 and 9 (16) 2 and 10 (17) 6 and 15 (18) 8 and 9
(19) 20 and 50 (20) 9 and 10

USING LCM WITH FRACTIONS

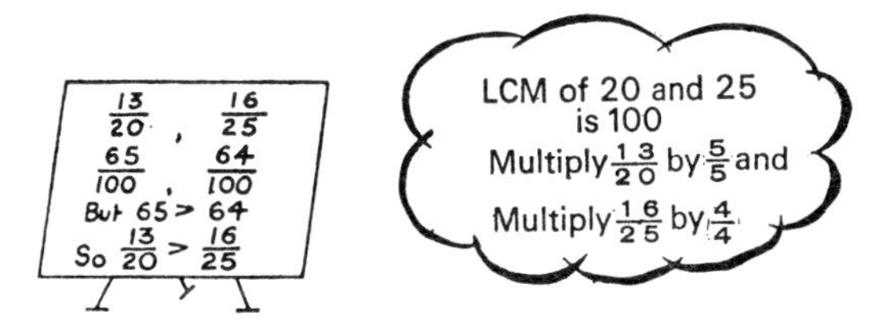

Write the correct symbol . . . (<, = or >) . . . between each pair of fractions using the LCM as on Abe's board:

54

(1) $\frac{1}{2}, \frac{1}{3}$ (2) $\frac{1}{2}, \frac{3}{5}$ (3) $\frac{2}{3}, \frac{3}{5}$ (4) $\frac{1}{3}, \frac{1}{4}$ (5) $\frac{1}{4}, \frac{2}{5}$
(6) $\frac{1}{4}, \frac{1}{6}$ (7) $\frac{3}{6}, \frac{5}{10}$ (8) $\frac{5}{6}, \frac{7}{8}$ (9) $\frac{4}{6}, \frac{6}{9}$ (10) $\frac{1}{4}, \frac{2}{9}$
(11) $\frac{9}{10}, \frac{5}{6}$ (12) $\frac{1}{3}, \frac{3}{8}$ (13) $\frac{9}{12}, \frac{6}{8}$ (14) $\frac{5}{8}, \frac{3}{5}$ (15) $\frac{2}{10}, \frac{3}{15}$
(16) $\frac{10}{15}, \frac{4}{6}$ (17) $\frac{23}{30}, \frac{17}{20}$ (18) $\frac{8}{20}, \frac{10}{25}$ (19) $\frac{3}{10}, \frac{4}{9}$ (20) $\frac{31}{50}, \frac{11}{20}$

Fractions — Practice

Find the missing numerators and denominators:

55
(1) $\frac{2}{3} = \frac{4}{\ }$ (2) $\frac{3}{4} = \frac{9}{\ }$ (3) $\frac{2}{5} = \frac{\ }{20}$ (4) $\frac{4}{7} = \frac{\ }{63}$
(5) $\frac{5}{8} = \frac{35}{\ }$ (6) $\frac{10}{15} = \frac{2}{\ }$ (7) $\frac{6}{24} = \frac{\ }{4}$ (8) $\frac{10}{50} = \frac{1}{\ }$
(9) $\frac{9}{21} = \frac{3}{\ }$ (10) $\frac{48}{54} = \frac{\ }{9}$

Bring these fractions to their lowest terms:

56
(1) $\frac{2}{10}$ (2) $\frac{9}{18}$ (3) $\frac{9}{12}$ (4) $\frac{25}{35}$ (5) $\frac{7}{21}$ (6) $\frac{70}{90}$
(7) $\frac{12}{21}$ (8) $\frac{24}{30}$ (9) $\frac{42}{56}$ (10) $\frac{12}{48}$ (11) $\frac{80}{100}$ (12) $\frac{63}{72}$

Change to mixed numbers:

57
(1) $\frac{8}{3}$ (2) $\frac{5}{4}$ (3) $\frac{9}{7}$ (4) $\frac{5}{2}$ (5) $\frac{12}{5}$ (6) $\frac{21}{4}$
(7) $\frac{47}{9}$ (8) $\frac{73}{7}$

Change to improper fractions:

(9) $1\frac{2}{3}$ (10) $1\frac{4}{7}$ (11) $1\frac{5}{8}$ (12) $2\frac{2}{9}$ (13) $3\frac{3}{4}$ (14) $5\frac{1}{7}$
(15) $6\frac{3}{8}$ (16) $9\frac{7}{10}$

Find the value of:

58
(1) $\frac{1}{4}$ of 36 (2) $\frac{1}{7}$ of 56 (3) $\frac{2}{3}$ of 24 (4) $\frac{4}{5}$ of 35
(5) $\frac{5}{9}$ of 27 (6) $\frac{1}{5}$ of 40m (7) $\frac{1}{6}$ of 48 books
(8) $\frac{2}{9}$ of 45 litres (9) $\frac{3}{8}$ of £72 (10) $\frac{7}{10}$ of 1 hour

EQUIVALENT FRACTIONS

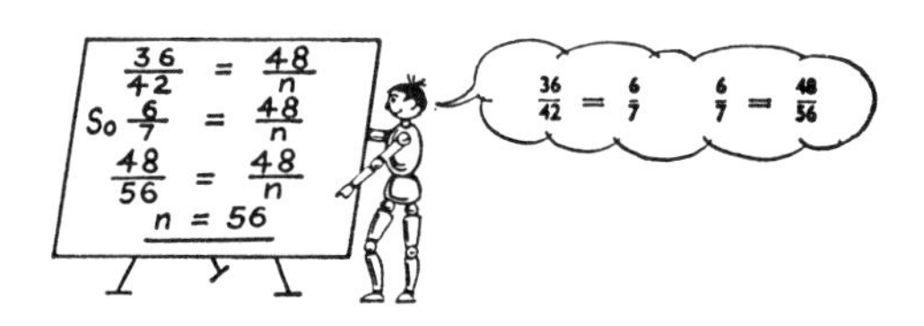

Find the replacement value of each letter:

59
(1) $\frac{1}{4} = \frac{a}{28}$ (2) $\frac{3}{5} = \frac{9}{b}$ (3) $1 = \frac{c}{6}$ (4) $\frac{18}{d} = \frac{2}{3}$
(5) $\frac{4}{8} = \frac{e}{2}$ (6) $\frac{2}{f} = \frac{16}{40}$ (7) $\frac{10}{30} = \frac{1}{g}$ (8) $\frac{10}{16} = \frac{h}{8}$
(9) $\frac{3}{10} = \frac{\ }{50}$ (10) $\frac{24}{42} = \frac{4}{j}$ (11) $\frac{k}{4} = \frac{24}{32}$ (12) $\frac{72}{p} = \frac{8}{9}$
(13) $\frac{0.3}{0.8} = \frac{6}{m}$ (14) $\frac{4}{18} = \frac{n}{27}$ (15) $\frac{6}{9} = \frac{p}{21}$ (16) $\frac{3}{15} = \frac{2}{p}$
(17) $\frac{20}{r} = \frac{16}{20}$ (18) $\frac{18}{42} = \frac{30}{s}$ (19) $\frac{t}{72} = \frac{28}{63}$ (20) $\frac{35}{40} = \frac{u}{32}$

FRACTIONS – PRACTICE

60

Find the answers to:

$2\frac{1}{4} + \frac{5}{6} - \frac{1}{2}$
$= 2\frac{3}{12} + \frac{10}{12} - \frac{6}{12}$
$= 2\frac{7}{12}$

(1) $\frac{2}{7} + \frac{4}{7}$ (2) $\frac{3}{5} - \frac{1}{5}$ (3) $\frac{3}{8} - \frac{1}{8}$
(4) $\frac{7}{10} - \frac{3}{10}$ (5) $\frac{2}{5} + \frac{1}{10}$ (6) $\frac{2}{3} - \frac{1}{6}$
(7) $\frac{1}{4} + \frac{1}{6}$ (8) $1 - \frac{3}{4}$ (9) $\frac{2}{9} + \frac{4}{9}$
(10) $3\frac{1}{4} - 2\frac{1}{8}$ (11) $\frac{3}{8} + 2\frac{1}{8}$ (12) $5\frac{1}{2} - 2\frac{3}{10}$ (13) $2\frac{1}{4} + \frac{3}{4}$
(14) $2\frac{3}{4} - 1\frac{2}{3}$ (15) $\frac{3}{8} + \frac{3}{4}$ (16) $4 - \frac{3}{4}$ (17) $\frac{5}{8} + \frac{2}{3} + \frac{1}{2}$
(18) $2\frac{2}{3} - 1\frac{3}{4}$ (19) $1\frac{1}{3} + 2\frac{1}{2} + 1\frac{5}{6}$
(20) $5\frac{1}{2} - 2\frac{3}{5}$ (21) $3\frac{3}{4} + \frac{1}{8} - 1\frac{1}{2}$

61

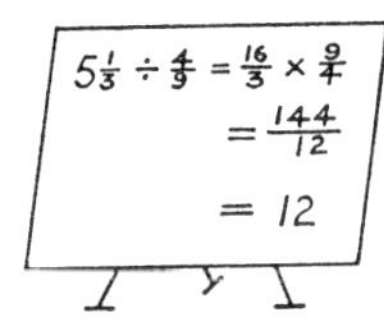

Find the answers to:

(1) $\frac{1}{3} \times 9$ (2) $\frac{1}{4} \times 5$ (3) $\frac{3}{5} \times 3$
(4) $\frac{5}{6} \times 2$ (5) $\frac{1}{3}$ of $\frac{5}{6}$ (6) $\frac{3}{5} \times \frac{2}{9}$
(7) $\frac{1}{4} \div 5$ (8) $\frac{2}{5} \div 7$ (9) $\frac{3}{8} \div 6$
(10) $5 \div \frac{1}{3}$ (11) $2 \div \frac{3}{4}$ (12) $\frac{9}{10} \div 6$ (13) $\frac{3}{4} \times 1\frac{1}{2}$
(14) $1\frac{1}{10} \div 1\frac{2}{5}$ (15) $4\frac{1}{2} \times \frac{1}{9}$ (16) $\frac{4}{5} \div 1\frac{2}{3}$ (17) $3\frac{1}{3} \times 3\frac{1}{3}$
(18) $\frac{3}{4} \div \frac{5}{6}$ (19) $1\frac{1}{5} \times \frac{3}{4} \times 2\frac{1}{2}$
(20) $4\frac{2}{3} \div 1\frac{2}{5}$ (21) $\frac{2}{3} \times \frac{3}{4} \div \frac{4}{5}$

PROBLEMS

62

(1) Ruth bought a packet of sweets. She ate one-third of the sweets in the morning and one-quarter in the afternoon. What fraction of the sweets did she eat in the day?

(2) Find the value of $\frac{3}{4}$ of $5\frac{1}{3}$ litres.

(3) Mother spent $\frac{1}{5}$ of her money at the Dairy and $\frac{1}{2}$ of her money at the Supermarket. What fraction of her money did she spend? What fraction of her money had she left?

(4) How many pieces of cloth each $2\frac{1}{4}$ metres long can be cut from a length of cloth 18 metres long?

(5) Find the total height of 6 tins each $3\frac{2}{3}$ cm tall.

(6) If I cut $2\frac{1}{2}$ cm from a piece of string $7\frac{2}{10}$ cm long, what length remains?

(7) A rectangle is $4\frac{2}{3}$ cm long and $2\frac{1}{4}$ cm broad

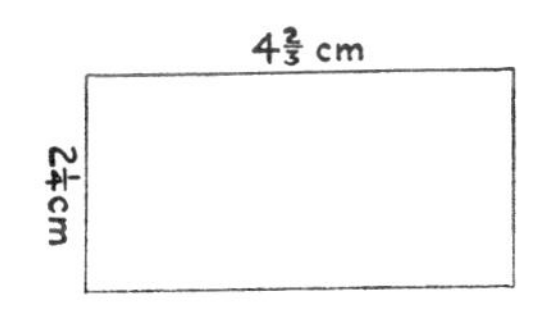

(a) Add the length of the rectangle to the breadth.

(b) Subtract the breadth of the rectangle from the length.

(c) Find the area of the rectangle in cm^2.

Fractions to Decimals

Abe says:

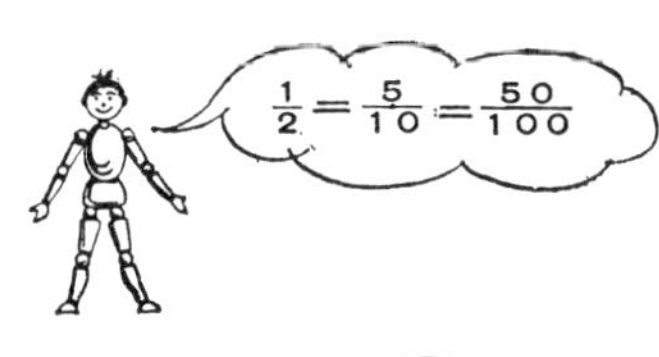

Abe thinks:

By finding equivalent fractions, write the following as decimals:

63

(1) $\frac{1}{2} = \frac{5}{10} = 0{\cdot}5$ (2) $\frac{3}{5} = \square = \square$

(3) $\frac{2}{5} = \square = \square$ (4) $\frac{4}{5} = \square = \square$

(5) $\frac{1}{4} = \frac{1 \times 25}{4 \times 25} = \frac{25}{100} = 0{\cdot}25$ (6) $\frac{3}{4} = \square = \square$

(7) $1\frac{1}{2} = 1 + \frac{1}{2} = 1 + \frac{5}{10} = 1{\cdot}5$ (8) $2\frac{3}{5} = \square + \square = \square$

(9) $2\frac{2}{5} = \square + \square = \square$ (10) $3\frac{3}{4} = \square + \square = \square$

Abe has a Problem

Smart Abe!

> If Abe continues he will get 0·3333333 . . . .
> He is happy with third decimal place approximation.

0·3333 . . . is written as 0·3 (Recurring decimal)

Express these fractions as decimals (to the 3rd decimal place):

64

(1) $\frac{1}{3}$ (2) $\frac{2}{3}$ (3) $0{\cdot}\dot{3}$ (4) $0{\cdot}\dot{6}$

(5) $1\frac{1}{3}$ (6) $4{\cdot}\dot{3}$ (7) $5{\cdot}\dot{3}\dot{1}$ (8) $7{\cdot}\dot{6}$

Preparing for Percentages

Here is part of a teacher's 'Mark Sheet':

NAME	English /100	History /25	Geography /25	Arithmetic /50	Science /20	Art /25
Robb, James	57	20	21	30	11	10
Small, Kay	72	21	22	39	12	11
Smith, Mary	46	12	9	20	16	14
Stephens, Ian	44	16	12	18	9	13
Todd, Karen	65	19	18	46	13	17

Abe compares James Robb's marks.

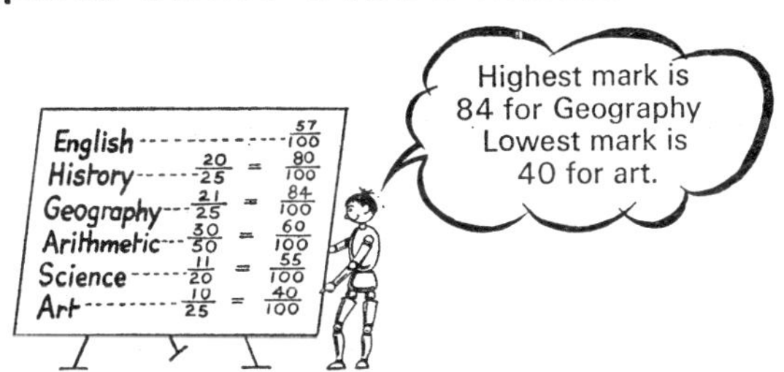

Why did Abe change all the marks to hundredths?

65

(1) Use the teacher's 'Mark Sheet' to find Kay Small's mark expressed as a fraction:
   (a) In History out of 25 (b) In History out of 100
   (c) In Geography out of 25 (d) In Geography out of 100
   (e) In Arithmetic out of 50 (f) In Arithmetic out of 100
   (g) In Science out of 20 (h) In Science out of 100
   (i) In Art out of 25 (j) In Art out of 100

(2) (a) In which subject did Kay score her highest mark out of 100?
   (b) In which subject did Kay score her lowest mark out of 100?

(3) (a) Compare Mary Smith's marks
   (b) What is her highest mark out of 100?

(4) (a) Compare Ian Stephen's marks
   (b) What is his lowest mark out of 100?

(5) (a) Compare Karen Todd's marks.
   (b) In which subjects did she score the same mark out of 100?

(6) What fraction of £1 is 17p?

(7) What fraction of £1 is 20p? Change the fraction to its lowest terms.

(8) What fraction of 1 metre is 69 cm?

(9) Find the fraction that 75 cm is of 1 metre in lowest terms.

PERCENTAGES

%

A *percentage* is a fraction with denominator 100.

$\frac{40}{100}$ of the squares were shaded.

$\frac{40}{100}$ may be written 40%.

So 40% of the squares were shaded.
40% is read as 40 per cent.

What percentage of the squares are blank?

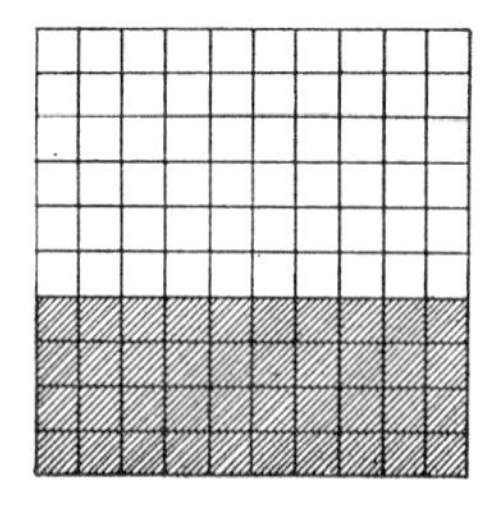

Write these percentages as fractions in lowest terms:

66

(1) 30% (2) 55% (3) 62% (4) 4% (5) 20%
(6) 35% (7) 14% (8) 60% (9) 25% (10) 56%
(11) 86% (12) 50% (13) 95% (14) 58% (15) 37%
(16) 75% (17) 72% (18) 80% (19) 69% (20) 100%

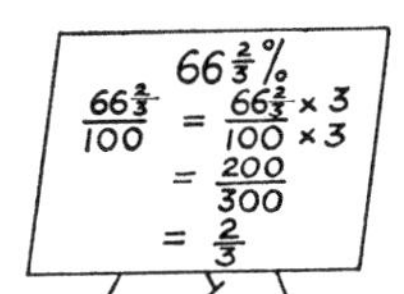

Express these percentages as fractions in lowest terms:

67

(1) $33\frac{1}{3}\%$ (2) $2\frac{1}{2}\%$ (3) $12\frac{1}{2}\%$ (4) $37\frac{1}{2}\%$ (5) $16\frac{2}{3}\%$
(6) $8\frac{1}{3}\%$ (7) $62\frac{1}{2}\%$ (8) $6\frac{1}{4}\%$ (9) $83\frac{1}{3}\%$ (10) $87\frac{1}{2}\%$

What is the total number of small squares?

What percentage of the small squares are shaded?

What percentage of the small squares are blank?

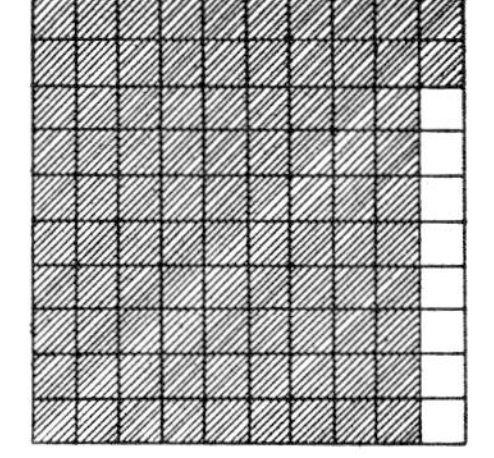

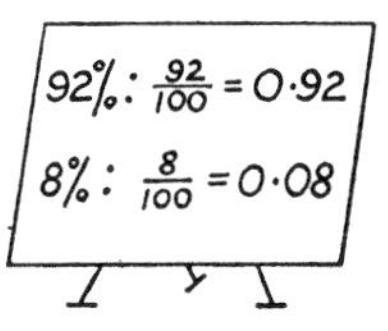

Write these percentages as decimals:

68

(1) 41% (2) 38% (3) 79% (4) 76% (5) 65%
(6) 56% (7) 70% (8) 83% (9) 11% (10) 9%
(11) 50% (12) 42% (13) 1% (14) 93% (15) 20%
(16) 5% (17) 80% (18) 16% (19) 6% (20) 37·5%

(21) Copy and complete this table:

Percentage	Fraction	Decimal	Percentage	Fraction	Decimal
10%			30%		
5%			70%		
20%			40%		
25%			60%		
50%			90%		
75%			80%		

MONEY AND LENGTH

69

(1) How many new pence in £1 ? (2) What fraction of £1 is 1p ?
(3) What percentage of £1 is 1p ? (4) What fraction of £1 is 3p ?
(5) What percentage of £1 is 3p ?
(6) What percentage of £1 is 10p ?
(7) How many cm are in 1 metre ?
(8) What fraction of 1 metre is 1 cm ?
(9) What percentage of 1 metre is 1 cm ?
(10) What fraction of 1 metre is 11 cm ?
(11) What percentage of 1 metre is 11 cm ?
(12) What percentage of 1 metre is 24 cm ?
(13) What percentage of £1 is 37p ?
(14) What percentage of £1 is 76p ?
(15) What percentage of £1 is 95p ?
(16) What percentage of 1m is 14 cm ?
(17) What percentage of 1m is 59 cm ?
(18) What percentage of 1m is 73 cm ?

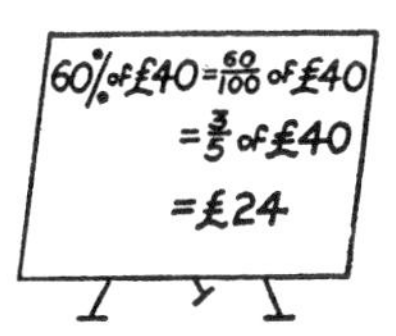

OR

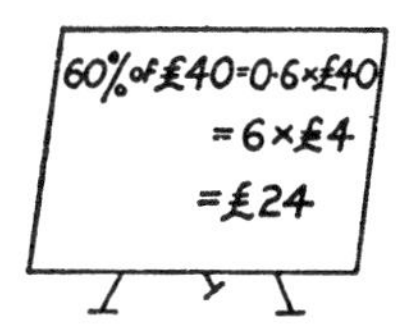

Find the answers to:

70

(1) 10% of 30p (2) 30% of 30p (3) 70% of 30p
(4) 90% of 30p (5) 20% of 15 metres (6) 40% of 15 m
(7) 60% of 15 m (8) 80% of 15 m (9) 25% of 80 kg
(10) 75% of 80 kg (11) 5% of £20 (12) 35% of £20
(13) 50% of 54 days (14) 30% of 90 oranges
(15) 70% of 180 eggs (16) 80% of 95 girls (17) 100% of 38 m
(18) 35% of £40 (19) 24% of 25 litres (20) 68% of 100 cc
(21) $12\frac{1}{2}$% of 32 boys (22) $37\frac{1}{2}$% of 56 cakes
(23) $33\frac{1}{3}$% of 24 pens (24) $66\frac{2}{3}$% of 18 discs (25) 20% of £3·45
(26) 10% of £16·30 (27) 30% of £18·20 (28) 80% of £9·25
(29) 90% of £2 (30) 40% of £24·35 (31) 70% of £3·60
(32) 6% of £5

FROM FRACTIONS AND DECIMALS TO PERCENTAGES

From a teacher's mark sheet, we saw earlier that John Todd scored these marks:

English ...	$\frac{65}{100}$		$= 65\%$
History ...	$\frac{19}{25}$	$= \frac{76}{100}$	$= 76\%$
Arithmetic ...	$\frac{46}{50}$	$= \frac{92}{100}$	$= 92\%$
Art	$\frac{13}{20}$	$= \frac{65}{100}$	$= 65\%$

> *Rule*
> To express a fraction as a percentage, find the equivalent fraction with denominator 100.

Abe uses two methods

$\frac{11}{20}$

Express these fractions as percentages:

71

(1) $\frac{1}{10}$ (2) $\frac{1}{5}$ (3) $\frac{1}{2}$ (4) $\frac{1}{4}$ (5) $\frac{3}{10}$
(6) $\frac{2}{5}$ (7) $\frac{3}{4}$ (8) $\frac{7}{10}$ (9) $\frac{3}{20}$ (10) $\frac{4}{5}$
(11) $\frac{1}{25}$ (12) $\frac{9}{10}$ (13) $\frac{3}{5}$ (14) $\frac{6}{25}$ (15) $\frac{19}{20}$
(16) $\frac{17}{50}$ (17) $\frac{19}{25}$ (18) $\frac{1}{8}$ (19) $\frac{5}{8}$ (20) $\frac{1}{3}$

$0{\cdot}8 = \frac{8}{10} = \frac{80}{100}$ or 80%
$0{\cdot}47 = \frac{47}{100}$ or 47%
$0{\cdot}765 = \frac{765}{1000} = \frac{76{\cdot}5}{100}$ or 76·5%

EASY!

Write these decimals as percentages:

72

(1) 0·56 (2) 0·92 (3) 0·7 (4) 0·63 (5) 0·2
(6) 0·62 (7) 0·5 (8) 0·94 (9) 0·16 (10) 0·43
(11) 0·03 (12) 0·1 (13) 0·08 (14) 0·9 (15) 1·00
(16) 0·125 (17) 0·54 (18) 0·725 (19) 0·05 (20) 0·875

(21) Copy and complete the following table:

Fraction	Decimal	Percentage	Fraction	Decimal	Percentage
$\frac{1}{5}$			$\frac{3}{10}$		
	0·1				60%
	0·4			0·25	
$\frac{1}{2}$			$\frac{4}{5}$		
		70%			75%
$\frac{1}{3}$	0·333	$33{\cdot}\dot{3}\%$			$66{\cdot}\dot{6}\%$

PROBLEMS ON PERCENTAGES

73

(1) Jean spends 40% of each day in bed.
What percentage of each day is Jean out of bed?

(2) Tom scored 12 marks out of 15 in a test. What percentage is this?

(3) There are 21 boys in a class of 35 pupils.
(a) What percentage of the class are boys? (b) What percentage are girls?

(4) 10% of 320 pupils in a school were absent on a certain day.
(a) How many pupils were absent?
(b) How many pupils were present?
(c) What percentage of the pupils were present?

(5) During a sale, a shop offered dresses at 25% less than the marked price. Mother bought a dress marked at £8.
(a) How much did she save? (b) What did the dress cost her?

(6) Father bought a shirt marked at £3·50. 20% was taken off this price.
(a) How much did he save?
(b) What did the shirt cost him?

(7) The length of a building was 60 metres. An extension was built which added 15% to the length.
(a) What is the length of the extension?
(b) What is the new length of the building?

(8) Nicola has 20p pocket money each week. If she spends 100% of her pocket money, how much does she save?

(9) A school team played 12 matches and lost 3 of them. What percentage of the total number of matches did the team lose?

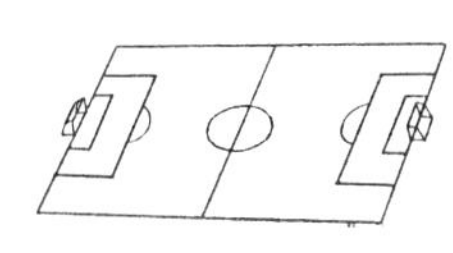

(10) There were 440 litres of petrol in a tank. 45% of the petrol was drained off. How much petrol remained in the tank?

(11) A father is 85% heavier than his son who weighs 40 kg. Find the weight of the father.

(12) A television set is marked at £84. Find the amount paid if 30% is taken off the price.

Progress Checks

PROGRESS CHECK 4

(1) Round off to the nearest ten: 58.
(2) Round off to the nearest centimetre: 49·8 cm.
(3) Multiply 8·24 by 3·16.
(4) Express £13·823 to the nearest penny.
(5) Find the HCF of 40 and 60.
(6) Find the value of: $\frac{4}{9}$ of £72.
(7) Find the answer to: $5\frac{3}{4} + 1\frac{2}{3}$.
(8) Express $3\frac{3}{5}$ as a decimal.
(9) Express 45% as a fraction in lowest terms.
(10) Express $\frac{17}{20}$ as a percentage.

PROGRESS CHECK 5

(1) Round off to the nearest hundred: 247.
(2) Round off to the nearest metre: 15·15 m.
(3) Multiply 9·16 by 3·24.
(4) Express 41·793 m to the nearest cm.
(5) Find the LCM of 8 and 12.
(6) Find the value of $\frac{10}{7}$ of 42 metres.
(7) Find the answer to: $3\frac{1}{3} \div \frac{3}{10}$.
(8) Express $1\frac{2}{3}$ as a decimal.
(9) Express 40% as a fraction in lowest terms.
(10) Express 0·24 as a percentage.

PROGRESS CHECK 6

(1) Father spent $2\frac{1}{2}$ hours working in the garden before lunch and $1\frac{3}{4}$ hours in the garden after lunch. What is the total time he spent in the garden?
(2) Having read $\frac{4}{9}$ of a book, what fraction of the book remains to be read?
(3) What fraction of an hour is 25 minutes?
(4) Find the weight of 84 sacks of flour if each sack weighs 42·6 kg.
(5) At netball shooting, Jean scored 18 times out of 24 throws and Mary scored 10 times out of 16 throws. Who is the better 'shooter', Jean or Mary?
(6) The page of a book is 9·3 cm long and 6·2 cm broad. Find the area of the page in cm^2 and write your answer to the first decimal place.
(7) Find the value of 18% of £2.
(8) Jean scored 13 marks out of 20 in a class test. What percentage is this?
(9) Father bought a television set which was priced at £75·40. 20% was taken off the price. What did the set cost him?
(10) Each day Abe spends $\frac{1}{12}$ of the time 'eating', $\frac{5}{12}$ of the time 'sleeping', $\frac{1}{6}$ of the time 'relaxing' and the rest of the time 'working'. What fraction of the day does Abe work?

Averages

Jack claimed he had better marks than Jean, so Abe decided he would act as judge and settle the dispute. He drew up a table:

Jack's Marks	43	39	46	Absent	37	49	35	38
Jean's Marks	40	45	41	37	40	43	36	38

Jack's total = 287 ... ... for 7 tests.
Jean's total = 320 ... ... for 8 tests.

Abe said, Jack scored marks at the rate of $\frac{287}{7}$ or 41 per test and Jean scored marks at the rate of $\frac{320}{8}$ or 40 per test.

So Jack is the winner!

Can you see why Jack was chosen as the winner?

We say, Jack's *AVERAGE* mark is 41,
Jean's *AVERAGE* mark is 40.

Note: $$\text{AVERAGE} = \frac{\text{THE SUM OF THE ITEMS}}{\text{THE NUMBER OF ITEMS}}$$

Find the average of the following:

74

(1) 6, 7, 8, 9, 10, 11, 12.
(2) 12, 11, 10, 10, 11, 12.
(3) 35, 24, 29, 37, 40.
(4) 93, 105, 99, 90, 112, 89.
(5) 7·3, 6·9, 8·4, 7·8.
(6) £14·25, £13·93, £15·30, £12·08.
(7) 3·57 m, 3·68 m, 3·80 m, 3·55 m.
(8) 10·2 l, 9·7 l, 10·5 l, 8·9 l, 9·9 l.
(9) 40·256 km, 39·815 km, 39·974 km.
(10) 0·535 kg, 0·587 kg, 0·561 kg.
(11) 65%, 57%, 92%, 77%, 54%.
(12) $1\frac{1}{4}$ hours, 55 m, $\frac{3}{4}$ h, 1 h 20 m.

PROBLEMS

75

(1) The weights of six pupils are, 29 kg, 37 kg, 33 kg, 35 kg, 43 kg, 39 kg. Find their average weight.

(2) The heights of five pupils are, 125 cm, 135 cm, 147 cm, 155 cm 143 cm. Find their average height.

(3) The numbers of books borrowed daily from a library are: Monday 135, Tuesday 110, Thursday 97, Friday 156, Saturday 192. Find the average number of books borrowed daily.

(4) Money collected from 10 classes for the school outing was: £3·10, £3·55, £2·80, £4·25, £3·95, £4·65, £3·65, £4·05, £3·00, £2·50. Find the average amount collected.

(5) Find the average number of words per line in the assignments below.

(6) Find the average number of letters per word in this short sentence.

(7) Find the average age of 5 pupils whose ages are, 11 years 3 months, 10 years 11 months, 11 years 4 months, 10 years 9 months, 11 years 2 months.

(8) On an 8-hour journey a car travels each hour, 64 km, 72 km, 56 km, 0 km, 61 km, 74 km, 45 km, 56 km. Find the average distance travelled per hour.

ASSIGNMENTS

(1) Find the time devoted to musical programmes on BBC TV each evening for a week between 1800 hours and close down. Find the average daily time allotted to such programmes.

(2) Find in centimetres the average height of (a) boys and (b) girls in your class. Are the girls on average taller than the boys?

(3) Find the "Editorial" column in a newspaper. Choose any ten lines and find the average number of words per line.
Estimate or guess the number lines in the column and thus find roughly the number of words in the editorial.

More Decimal Multiplication

Multiply 4·25 by 3·25 and give answer correct to 2nd decimal place.

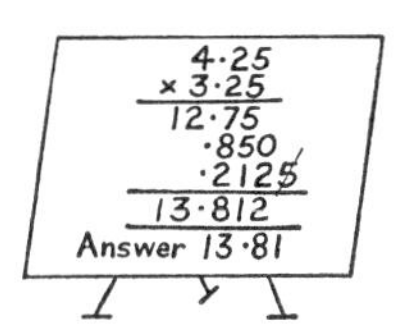

Multiplication:

Give answers correct to 2nd decimal place:

76

(1) $2{\cdot}12 \times 3{\cdot}23$ (2) $4{\cdot}23 \times 3{\cdot}25$ (3) $5{\cdot}16 \times 2{\cdot}34$

(4) $6{\cdot}32 \times 4{\cdot}14$ (5) $6{\cdot}42 \times 5{\cdot}32$ (6) $7{\cdot}18 \times 6{\cdot}28$

(7) $8{\cdot}25 \times 6{\cdot}82$ (8) $5{\cdot}84 \times 9{\cdot}27$ (9) $9{\cdot}16 \times 7{\cdot}83$

(10) $7{\cdot}68 \times 8{\cdot}34$

ABE'S PUZZLERS

1 g = 0·001 kg
1 p = £0·01
1 cm = 0·01 m

77

(1) Find to the third place of decimals the area of a rectangle 2·25 m long and 3·50 m broad.

(2) Find to the nearest penny the cost of 2·75 kg of potatoes at 7·5 p/kg.

(3) What is the cost of 8·75 units of electricity at 2·65p for one unit? Give your answer to the nearest whole penny.

(4) What is *?
$(2{\cdot}5 \times 1{\cdot}5) + (3{\cdot}25 + 2{\cdot}75) = *$

(5) What is the area of the floor of a room 4·25 m by 3·75 m. Give your answer to the second place of decimals.

Money, Weight, Length, Liquid Measure

ABE'S MONEY PROBLEMS

78

(1) If 6 litres of petrol costs £1·86 what will be the cost of **1** litre of petrol?

(2) How much will 12 litres of petrol cost if the price for one litre is the same as in Question 1?

(3) 

A man has £10 and spends £4·34 in one shop and £3·25 in another. How much money has he left?

(4) 7 men are paid the same weekly wage. The total wage bill for the seven men is £126·70. How much money does each man receive?

(5) A calculating machine costs £42·25. What is the total cost of 36 such calculating machines.

(6) 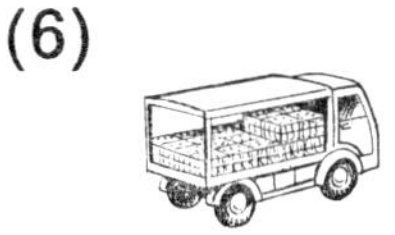
A lorry carries 40 crates of milk. Each crate holds 24 litres milk. One litre of milk costs 20p. How much money is the lorry load worth?

(7) Apples cost £3·25 per case. How much will you pay for 80 cases?

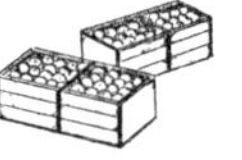

(8) A slide rule is valued at £1·52. What is the total value of 48 such slide rules?

(9) A woman buys three dresses at £1·25, £3·56 and £4·84. How much change does she get from £10?

(10) Find the replacement value for $\triangle$: $(£7{\cdot}85 - £4{\cdot}35) \times 12 = \triangle$.

WEIGHT-WATCHER ABE HAS PROBLEMS

1 kilogramme	is 1 kg
1 gramme	is 1 g

1 kg = 1000 g

1 g = 0·001 kg

79

(1) A box weighs 1·5 kg when empty and 7·5 kg when filled with apples. What is the total weight of apples in 24 such boxes?

(2)  Nine boys have a total weight of 540 kg. What is their average weight?

(3) Eight rugby forwards weigh 98·5 kg, 110 kg, 105·2 kg, 99·7 kg, 104·6 kg, 103·8 kg, 94·2 kg and 109·6 kg respectively. What is their average weight?

(4) A lorry, fully loaded, weighs 1 242 kg. What is the total weight of twenty-three similar lorries.

(5) A box of chocolates weighs 0·55 kg. What will 15 similar boxes weigh?

(6) A calculating machine weighs 3·8 kg and its container weighs 1·5 kg. How much will 12 such machines and their containers weigh, altogether?

(7) 250 g is the weight of a packet of tea. What is the total weight of twenty-four of these packets?

(8) Find the total number of kg in: (4·25 kg + 6·75 kg) + (8·35 kg + 7·65 kg).

(9) Find the answer to (7·95 kg + 2·83 kg) ÷ 7.

(10) A box weighs 2·35 kg and, when full, holds a weight of 18·85 kg of tins. How much will the box and tins together weigh if a weight of 16·38 kg of these tins is removed?

PROBLEMS – LENGTH

Cycle Track

80

(1) The perimeter of a cycle racing track is 400 m. How many times must a cyclist go round to complete 4 000 m?

(2) The circumference of a wheel is 3 m. How many times does the wheel turn in covering 4 800 metres?

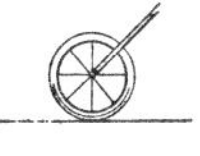

(3)

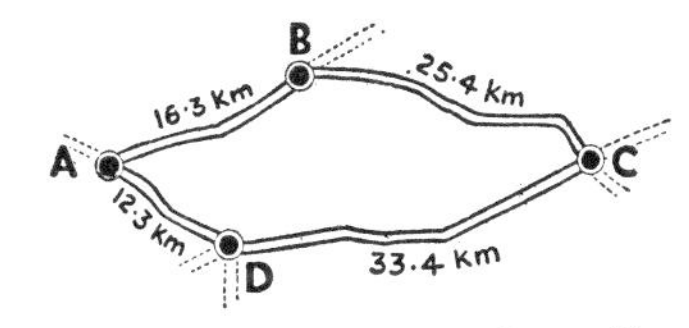

A, B, C and D are four towns and the distances between them are shown on the map.

Answer these questions from your map.

(a) How far is it from A to C via B?
(b) How far is it from A to C via D?
(c) Which is the shorter journey, (a) or (b)?
(d) What is the shorter route from B to D?

(4) A train is 200 metres long and a bridge, which it has to cross, is 400 metres long. How far will the engine travel, on crossing the bridge, before the whole train is clear of the bridge?

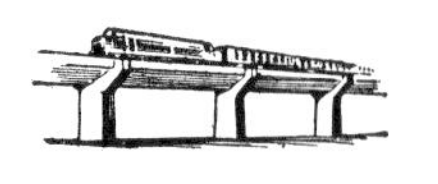

(5) A coil of wire 10 m long, has 120 pieces of wire, each 7·8 cm long, cut off its length. What length of wire is now in the coil?

(6) The average weekly distance travelled by a salesman is 320 kilometres. How far does he travel in a year (52 weeks)?

(7) A rectangular field is 53·6 metres long and 48·3 metres wide. How much fencing is required to surround the field if two gates 2·5 metres wide are to be made?

PROBLEMS – LIQUID MEASURE

8I

(1) A cow provides 4·3 litres of milk each day. How much milk would be provided in 7 days?

(2) A farmer collects 1 470 litres milk in one week. What is the daily average for that week?

(3) How many 2 litre bottles can be filled from a 1 000 litre container?

(4) A tank holds 41·5 litres petrol. How much petrol is there in 12 such tanks?

(5) A farm cottage is supplied with water from a 300 litre tank which is filled after 14 days. The cottage people use 20 litres each day. How much water will be in the tank when it is due to be filled?

(6) A dairy supplies 44 customers with 2 litres milk each day, 12 customers with 0·5 litres and 4 customers with 1 litre. At the end of one day there were 12 litres left. How much milk was there at the beginning of the day?

(7) Milk arrives at a shop in a container which holds 48·6 litres and is used to fill 6 milk cans, all of equal volume. How much milk will one can hold?

(8) A hose delivers 4 032 litres of water in 3 hours into a pond. If the flow of water is at a constant speed, how many litres does the hose deliver in 5 minutes?

(9) If a litre of a certain liquid weighs 875 grammes, find the weight of 15 litres of this liquid in kilogrammes.

(10) The flow of water into a cistern is 28·5 litres every minute. How much water flows into the cistern in 1·5 hours?

Division

Abe says:

3·6 means 3 units and six tenths 3·60 means 3 units and sixty hundredths

Abe is puzzled

Division:

82

(1) $42{\cdot}7 \div 14$ (2) $43{\cdot}2 \div 12$ (3) $64{\cdot}8 \div 16$ (4) $75{\cdot}9 \div 15$
(5) $90{\cdot}6 \div 15$ (6) $48{\cdot}8 \div 16$ (7) $70{\cdot}7 \div 14$ (8) $967{\cdot}2 \div 12$
(9) $42{\cdot}0 \div 35$ (10) $85{\cdot}0 \div 25$

State which of the following are false:

83

(1) $42{\cdot}7 \div 14 = 3{\cdot}5$
(2) $36{\cdot}12 \div 12 = 3{\cdot}04$
(3) $27{\cdot}9 \div 9 = 3{\cdot}00$
(4) $64{\cdot}1 \div 21 = 3{\cdot}1$
(5) $6{\cdot}51 \div 21 = 0{\cdot}31$
(6) $279 \div 9 = 31$
(7) $36{\cdot}00 \div 4 = 9{\cdot}00$
(8) $0{\cdot}36 \div 9 = 0{\cdot}4$
(9) $86{\cdot}1 \div 21 = 4{\cdot}0$
(10) $6 \div 15 = 0{\cdot}4$

From the sets below find in each case the member which gives the answer to the calculation:

84

(1) 65·7 ÷ 9 {0·73, 7·3, 73}
(2) 7·28 ÷ 8 {0·91, 9·1, 0·091}
(3) 11·0 ÷ 10 {0·11, 1·1, 110}
(4) 28·91 ÷ 7 {40·13, 4·13, 41·3}
(5) 72·48 ÷ 12 {6·04, 60·4, 6·4}
(6) 75·00 ÷ 15 {5, 85·0, 85·00}
(7) 54·40 ÷ 17 {3·2, 32, 3·20}
(8) 48·30 ÷ 23 {2·1, 2·10, 21}
(9) 214·2 ÷ 21 {1·2, 1·02, 10·2}
(10) 42·40 ÷ 20 {2·12, 21·2, 2·012}

DECIMALS

34·5 ÷ 15

Division:

85

(1) 38·4 ÷ 12	(2) 28·6 ÷ 13	(3) 43·4 ÷ 14
(4) 48·3 ÷ 23	(5) 69·3 ÷ 33	(6) 75·6 ÷ 36
(7) 50·4 ÷ 24	(8) 59·4 ÷ 27	(9) 96·1 ÷ 31
(10) 92·4 ÷ 84	(11) 85·8 ÷ 78	(12) 73·7 ÷ 67
(13) 15·6 ÷ 12	(14) 78·1 ÷ 71	(15) 108·9 ÷ 99

Division:

86

(1) 154·5 ÷ 15	(2) 173·4 ÷ 17	(3) 262·6 ÷ 26
(4) 343·4 ÷ 17	(5) 763·8 ÷ 38	(6) 886·2 ÷ 42
(7) 107·1 ÷ 51	(8) 798·38 ÷ 38	(9) 1085·4 ÷ 54

State which of the following sentences are true and which are false:

87

(1) 77·7 ÷ 7 = 1·11	(2) 8·4 ÷ 84 = 0·1
(3) 7·2 ÷ 6 = 0·12	(4) 35 ÷ 70 = 0·5
(5) 26·13 ÷ 13 = 2·01	(6) 15·6 ÷ 12 = 1·03
(7) 28·6 ÷ 13 = 20·2	(8) 17·3 ÷ 2 = 8·65
(9) 34·5 ÷ 15 = 23	(10) 16·16 ÷ 16 = 1·1

PROBLEMS

88

(1) Twelve boxes of equal weight have a total weight of thirty-eight point four kilogrammes. What is the weight of one box?

(2) The total marks scored by twenty-five pupils is 707. What is their average mark?

(3) Twenty-four boys save £50·40 and they decide to share equally. What amount of money does each boy receive?

(4) In a skating contest a skater was awarded by five judges 27·75 points, 29·50 points, 22·25 points, 25·50 points and 24·50 points. What was the average mark awarded?

(5) The account for a birthday party is £75·60. Twelve people are paying equal shares. What will each person pay?

(6) A colour TV set costs £360. Two years are allowed to pay and 20% is added as a service charge. How much will a buyer pay each month?

(7) Thirty-six tins of beans weigh 16·20 kg. What is the average weight of one tin?

(8) A car travels 772·8 kilometres on 24 litres of petrol. How many kilometres would the same car travel on 1 litre of petrol?

(9) Forty-eight sacks of potatoes weigh 585·6 kg. What is the average weight of one sack?

(10) The cost of 10 litres of petrol is 55p. How much will one litre of petrol cost?

Division by Decimals

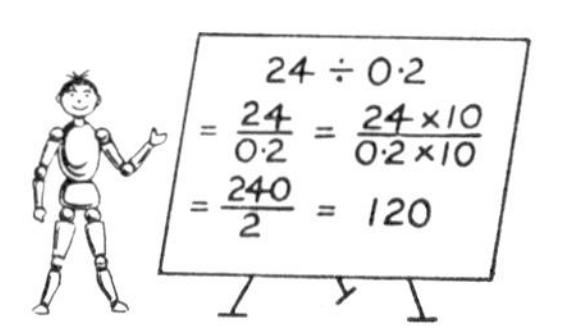

Again 36·3 ÷ 0·3

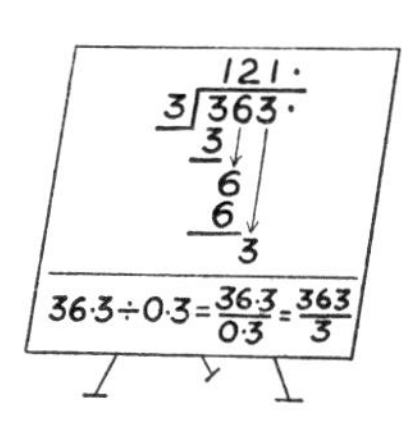

> To divide by a decimal find the equivalent fraction which has a whole number as Denominator.

89 Express as equivalent fractions with whole numbers as denominators:

(1) $\frac{7\cdot2}{0\cdot6}$ (2) $\frac{4\cdot8}{0\cdot4}$ (3) $\frac{8\cdot4}{0\cdot7}$ (4) $\frac{9\cdot6}{0\cdot8}$ (5) $\frac{3\cdot6}{0\cdot2}$

(6) $\frac{7\cdot2}{0\cdot3}$ (7) $\frac{9\cdot6}{0\cdot4}$ (8) $\frac{10\cdot8}{0\cdot9}$ (9) $\frac{2\cdot5}{0\cdot5}$ (10) $\frac{5\cdot4}{0\cdot6}$

90 Division:

(1) 7·2 ÷ 0·6 (2) 4·8 ÷ 0·4 (3) 8·4 ÷ 0·7
(4) 9·6 ÷ 0·8 (5) 3·6 ÷ 0·2 (6) 7·2 ÷ 0·3
(7) 9·6 ÷ 0·4 (8) 10·8 ÷ 0·9 (9) 2·5 ÷ 0·5
(10) 5·4 ÷ 0·6 (11) 9·64 ÷ 0·4 (12) 7·29 ÷ 0·3

Find the replacement for □:

(13) 1·8 ÷ 0·6 = □ (14) 0·18 ÷ 0·3 = □
(15) 2·0 ÷ 0·5 = □ (16) 2·80 ÷ 0·7 = □
(17) 44·4 ÷ 0·4 = □ (18) 4·44 ÷ 0·4 = □
(19) 0·7 ÷ 0·7 = □ (20) 3·69 ÷ 0·9 = □
(21) 0·12 ÷ 0·4 = □ (22) 16·48 ÷ 0·04 = □

Abe thinks again

$0{\cdot}07 \times 100$

$\times$ 100: MOVE digits TWO places to LEFT

Abe uses his abacus

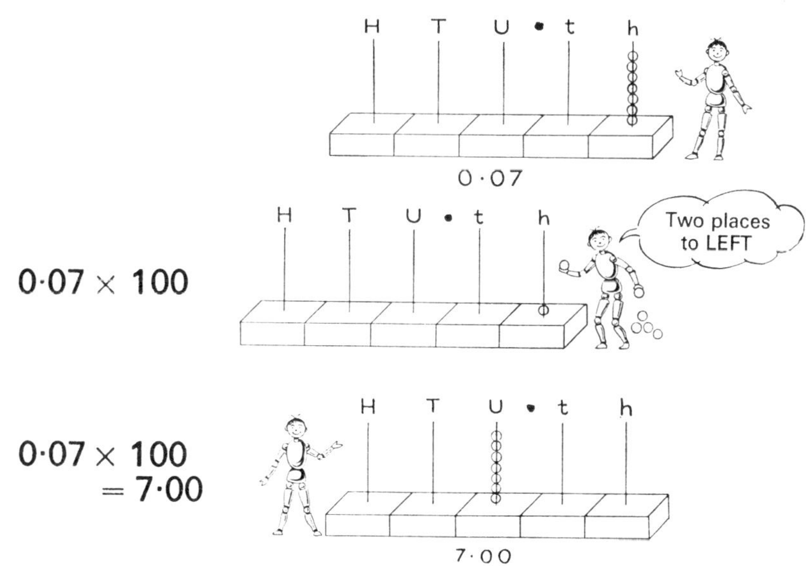

$0{\cdot}07 \times 100$

$0{\cdot}07 \times 100$
$= 7{\cdot}00$

Multiplication:

91

(1) $7{\cdot}47 \times 100$	(2) $2{\cdot}34 \times 100$	(3) $0{\cdot}08 \times 100$
(4) $1{\cdot}40 \times 100$	(5) $7{\cdot}23 \times 100$	(6) $8{\cdot}91 \times 100$
(7) $0{\cdot}03 \times 100$	(8) $2{\cdot}64 \times 100$	(9) $4{\cdot}06 \times 100$
(10) $2{\cdot}213 \times 100$	(11) $4{\cdot}269 \times 100$	(12) $7{\cdot}256 \times 100$

Find the equivalent fractions, with whole number denominator:

92

(1) $\frac{2{\cdot}44}{0{\cdot}04}$	(2) $\frac{6{\cdot}48}{0{\cdot}08}$	(3) $\frac{9{\cdot}54}{0{\cdot}02}$
(4) $\frac{5{\cdot}55}{0{\cdot}05}$	(5) $\frac{4{\cdot}97}{0{\cdot}07}$	(6) $\frac{64{\cdot}46}{0{\cdot}06}$
(7) $\frac{5{\cdot}55}{0{\cdot}11}$	(8) $\frac{28{\cdot}14}{0{\cdot}14}$	(9) $\frac{32{\cdot}48}{0{\cdot}16}$
(10) $\frac{0{\cdot}5555}{0{\cdot}11}$	(11) $\frac{0{\cdot}2814}{0{\cdot}14}$	(12) $\frac{0{\cdot}3248}{0{\cdot}16}$

Division:

93

(1) $2{\cdot}44 \div 0{\cdot}04$	(2) $6{\cdot}48 \div 0{\cdot}08$	(3) $9{\cdot}54 \div 0{\cdot}02$
(4) $65{\cdot}55 \div 0{\cdot}05$	(5) $14{\cdot}28 \div 0{\cdot}07$	(6) $42{\cdot}60 \div 0{\cdot}06$
(7) $5{\cdot}555 \div 0{\cdot}11$	(8) $64{\cdot}46 \div 0{\cdot}04$	(9) $0{\cdot}3248 \div 0{\cdot}16$

MORE DIVISION – DECIMAL

$56{\cdot}14 \div 1{\cdot}4$

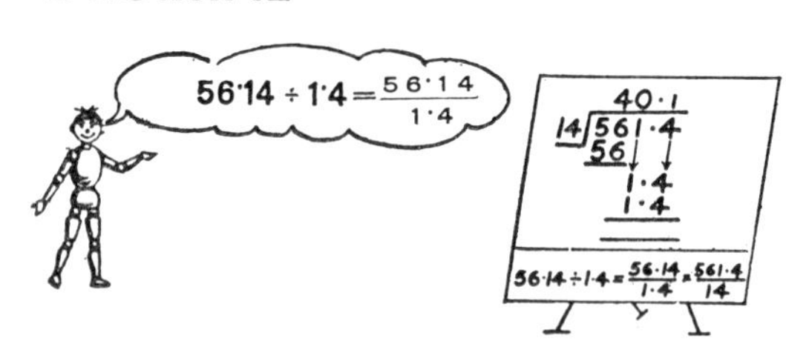

94 Write down in each case, the equivalent fraction with a whole number denominator:

(1) $\frac{96{\cdot}88}{0{\cdot}8}$ (2) $\frac{46{\cdot}65}{0{\cdot}15}$ (3) $\frac{692{\cdot}3}{2{\cdot}3}$ (4) $\frac{9{\cdot}495}{4{\cdot}5}$

(5) $\frac{7{\cdot}56}{0{\cdot}36}$ (6) $\frac{139{\cdot}5}{4{\cdot}5}$ (7) $\frac{117{\cdot}6}{0{\cdot}56}$ (8) $\frac{61{\cdot}61}{0{\cdot}61}$

Complete these equivalent fractions by finding the replacement for □ :

(9) $\frac{81{\cdot}18}{0{\cdot}09} = \frac{8\,118}{\square}$ (10) $\frac{68{\cdot}34}{3{\cdot}4} = \frac{\square}{34}$

(11) $\frac{0{\cdot}0648}{3{\cdot}24} = \frac{\square}{324}$ (12) $\frac{568{\cdot}508}{2{\cdot}54} = \frac{\square}{254}$

Division:

95

(1) $96{\cdot}88 \div 0{\cdot}8$ (2) $46{\cdot}65 \div 0{\cdot}15$ (3) $692{\cdot}3 \div 2{\cdot}3$
(4) $9{\cdot}495 \div 4{\cdot}5$ (5) $75{\cdot}6 \div 3{\cdot}6$ (6) $81{\cdot}18 \div 0{\cdot}09$
(7) $68{\cdot}34 \div 3{\cdot}4$ (8) $139{\cdot}5 \div 4{\cdot}5$ (9) $48{\cdot}12 \div 0{\cdot}12$
(10) $4{\cdot}812 \div 0{\cdot}12$ (11) $6{\cdot}432 \div 1{\cdot}6$ (12) $0{\cdot}6432 \div 0{\cdot}1$

From the sets below find in each case the member which is the replacement for □ :

96

(1) $2{\cdot}8 \div 1{\cdot}4 = \square$ $\{0{\cdot}2, 2, 0{\cdot}02\}$
(2) $0{\cdot}69 \div 2{\cdot}3 = \square$ $\{3{\cdot}0, 0{\cdot}3, 3\}$
(3) $7{\cdot}56 \div 0{\cdot}36 = \square$ $\{2{\cdot}1, 21{\cdot}0, 0{\cdot}21\}$
(4) $106{\cdot}53 \div 5{\cdot}3 = \square$ $\{2{\cdot}01, 20{\cdot}1, 21\}$
(5) $84{\cdot}42 \div 4{\cdot}2 = \square$ $\{2{\cdot}1, 2{\cdot}01, 20{\cdot}1\}$
(6) $7{\cdot}296 \div 2{\cdot}4 = \square$ $\{0{\cdot}304, 3{\cdot}04, 30{\cdot}4\}$
(7) $51{\cdot}34 \div 0{\cdot}17 = \square$ $\{302, 3{\cdot}02, 30{\cdot}2\}$
(8) $15{\cdot}351 \div 5{\cdot}1 = \square$ $\{30{\cdot}1, 3{\cdot}01, 3{\cdot}1\}$
(9) $99{\cdot}66 \div 0{\cdot}33 = \square$ $\{3{\cdot}02, 30{\cdot}2, 302\}$
(10) $6{\cdot}432 \div 1{\cdot}6 = \square$ $\{4{\cdot}02, 402, 42\}$

ABE'S DECIMAL MIXTURES

97

(1) In a class of 30 pupils the examination average mark was 54·2. What was the total of marks scored by the class?

(2) A man runs 100 metres in 10·2 seconds. If his speed is constant how long would he take to run 60 metres?

(3) Find the cost of 2·5 metres of cloth at £1·25 for one metre.

(4) A woman bought 3·75 kilogrammes of butter at 36p per kilogramme. How much did she pay altogether?

(5) Thirty-five sheep were bought by a farmer for £253·75. What was the average price per sheep?

(6) A boy has £5·75 in his savings account. He deposits 60p one week and 40p the next week. If he withdraws £2·60 the following week how much money remains in his account?

(7) 1·5 kilogrammes of butter cost 75 new pence. How much will one kilogramme of butter cost?

(8) A car uses 24 litres of petrol for a journey. The petrol consumption is 12·12 kilometres to the litre. How far has the car travelled?

(9) Petrol costs 35p for four litres. What was the cost of petrol in Question 8?

(10) An electricity meter read 4 812 three months ago. It now reads 5 264. If the average price for one unit of electricity is 1·2p calculate the amount of money to be paid to the nearest penny.

Time and Distance—Speed

FINDING SPEEDS

Examples:

(1) If a man walks 8 kilometres in 2 hours, we say his average speed is $\frac{8}{2}$ or 4 kilometres per hour = 4 km/hour

(2) If a train travels 600 kilometres in 10 hours its average speed is 60 km/hour. ($\frac{600}{10} = 60$)

98 Copy and complete the following table:

Distance (km)	Time (hours)	Speed (km/hour)
40	5	$\frac{40}{5} = 8$
120	2	
120	4	
3 000	5	
500	2·5	
20	0·5	
31 386	2	

99 Find the answers to the following:

(1) A boy cycles 32 kilometres in 2 hours. What is his average speed in km/hour?

(2) A plane flies 3 500 kilometres in 4 hours. Find its average speed in km/hour.

(3) The journey by road from Glasgow to London takes 10 hours. The distance is 640 kilometres. What is the average speed?

(4) If a satellite travels 360 000 kilometres in 2 days, what is its average speed in km/hour?

(5) A train travels 312 kilometres. Its departure time is 2200 hours and arrival time 0100 hours next morning. Find its average speed in km/hour.

(6) A plane leaves London at 2000 hours and arrives at 1200 hours next day at Johannesburg, 12 600 kilometres away. If it has a 1 hour refuelling stop, find its average speed in km/hour.

(7) A sprinter runs 100 metres in 10 seconds. What is his average speed in km/hour? (1 000 m = 1 km.)

FINDING DISTANCES

NOTE: When we say "the car was travelling at 40 km/hour", we mean that *if* it travelled for one hour it *would* travel 40 kilometres. If it travelled for two hours it *would* go 80 kilometres.

Examples:

(1) If a boy walks at an average speed of 3 km/h, how far does he walk in 4 hours?
distance = 12 kilometres. ($4 \times 3 = 12$)

(2) If a plane travels at an average speed of 800 km/h, how far does it fly in 5 hours?
distance = 4 000 kilometres. ($800 \times 5 = 4\,000$)

100 Copy and complete the following table:

Distance (km)	Time (hours)	Speed (km/hour)
	1	60
	2	30
	3	90
	10	160
	5	160
	15	160
1 200	1·5	

Find the answers to the following:

101

(1) A car travels at an average speed of 64 km/hour for $3\frac{1}{2}$ hours. How far does it travel?

(2) A bus leaves Glasgow at 1000 hours and arrives in Inverness at 1700 hours. It stops for 1 hour on the way. Find the distance from Glasgow to Inverness if the average speed of the bus is 46 km/hour.

(3) A plane has enough fuel for $5\frac{3}{4}$ hours flying time at an average speed of 640 km/hour. How far can it travel?

(4) A racing car travels at 244·8 km/hour for one minute. How far does it travel?

(5) A car speedometer shows 48 km/hour for 10 minutes, then 84 km/hour for 20 minutes. How far has the car gone in this half hour?

Reading Graphs

STRAIGHT LINE GRAPHS

102 A car travels at a *CONSTANT* speed of 50 km/hour for 5 hours.

The table below shows the distance the car has travelled after 1, 2, 3, 4 and 5 hours.

This information is used to draw the graph.

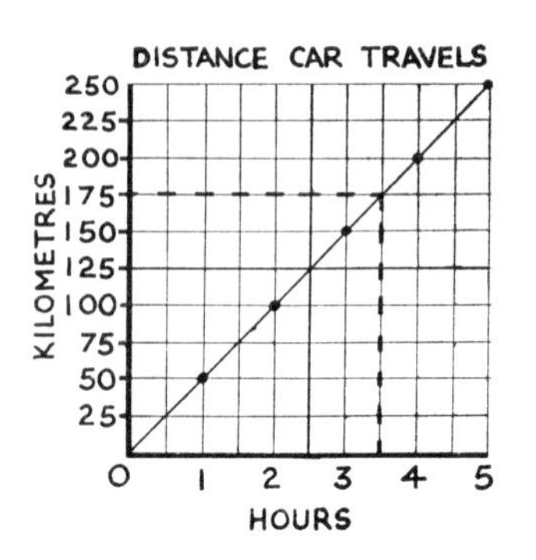

DISTANCE(km)	50	100	150	200	250
TIME (hours)	1	2	3	4	5

Answer the following questions from the graph:

(1) How far does the car travel in $4\frac{1}{2}$ hours?
(2) How far does the car travel in $2\frac{1}{2}$ hours?
(3) How far does the car travel each $\frac{1}{2}$ hour?
(4) How long does the car take to go 175 km?
(5) How long does the car take to go 75 km?

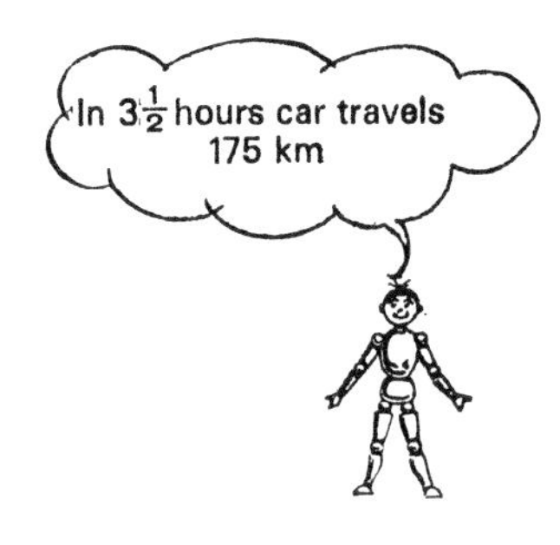

NOTE: All the points on the graph lie on a straight line. This is called a **STRAIGHT LINE** graph.

103 A runner keeps up a steady rate of 200 metres per minute.
NOTE: Three points are needed to draw a straight line graph.

TIME (minutes)	1	2	5
DISTANCE (Metres)	200	400	1000

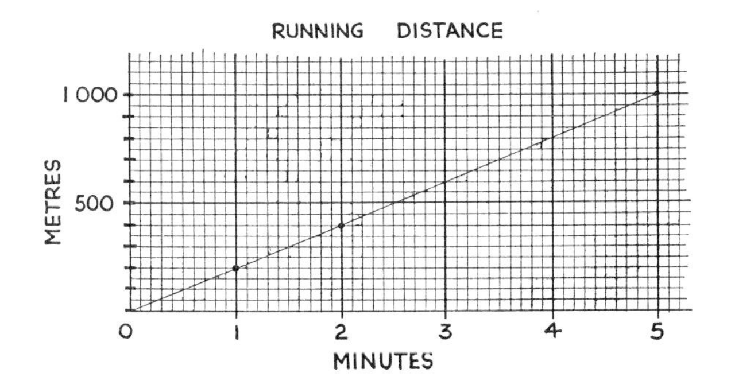

Answer these questions from the graph:

(1) How far does the runner go in (a) $\frac{1}{2}$ minute, (b) 3 minutes, (c) $1\frac{1}{2}$ minutes?

(2) How long does the runner take to go (a) 500 m (b) 700 m?

(3) What is the runner's constant speed in metres per minute?

104 One dinner ticket costs 40p.

NUMBER OF TICKETS	0	10	30
COST IN £	0	4	12

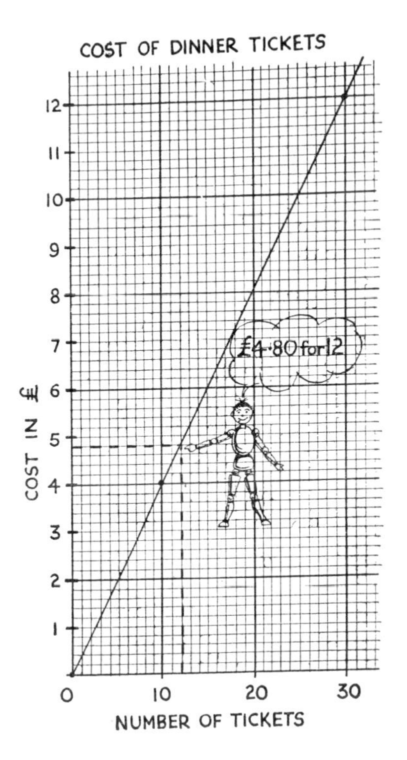

The graph shows the cost of up to 30 dinner tickets.

NOTES: (1) 0 tickets cost £0·00.
(2) 12 tickets cost £4·80.

From the graph answer the following questions:

(1) What is the cost of (a) 5 (b) 20 (c) 24 (d) 17 dinner tickets?

(2) How many tickets can be bought for (a) £6·00 (b) £8·40 (c) £11·20?

(3) How much change is there from
(a) £4·00 when 9 tickets are bought?
(b) £5·00 when 11 tickets are bought?

(4) How much should be paid for 42 dinner tickets?

Ratios

$\frac{4}{9}$ or 4 : 9

Ann bought a packet containing 9 sweets. She ate 4 sweets. What fraction of the sweets did she eat?

When *comparing numbers*, we use the term *RATIO*.

We say, "The ratio of the number of sweets bought to the number of sweets eaten is 9 to 4".

We write, Number of sweets bought: Number of sweets eaten is 9 : 4

or in fraction form, $\frac{\text{Number of sweets bought}}{\text{Number of sweets eaten}} = \frac{9}{4}$

NOTE: The ratio 9 : 4 or $\frac{9}{4}$ is read as "Nine to four".

The ratio, ... Sweets left : sweets bought is 5 : 9 or $\frac{5}{9}$

The ratio, ... Sweets eaten : sweets bought is 4 : 9 or $\frac{4}{9}$

105 In a multi-storey block of flats there live 26 men, 44 women, 14 boys and 16 girls.

For this block, find the total number of:

(1) People (2) Adults (3) Children (4) Males

Write down the value of the ratios:

(5) Men : People (6) Women : People (7) People : Boys
(8) People : Girls (9) Men : Boys (10) Women : Girls
(11) Girls : Boys (12) Adults : Men (13) Women : Adults
(14) People : Adults (15) Boys : Children (16) Girls : Children
(17) Children : People (18) Boys : Males (19) Men : Males
(20) Children : Adults

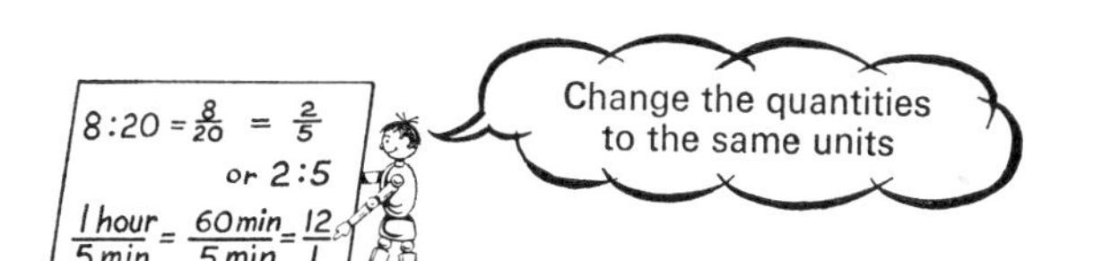

106 Write each ratio in its simplest form:

(1) $\frac{5}{10}$ (2) $\frac{3}{9}$ (3) $\frac{6}{8}$ (4) $\frac{12}{18}$

(5) $\frac{12}{20}$ (6) 14 : 20 (7) 21 : 35 (8) 16 : 40

(9) 50 : 40 (10) 32 : 28 (11) $\frac{10p}{20p}$ (12) $\frac{24 \text{ cm}}{32 \text{ cm}}$

(13) $\frac{32g}{20g}$ (14) 10 sec : 45 sec (15) 18 l : 33 l

(16) $\frac{10p}{£1}$ (17) $\frac{40 \text{ min}}{1 \text{ hour}}$ (18) $\frac{30 \text{ cm}}{1 \text{ metre}}$

(19) £1 : 45p (20) 40 cm : 2 m

(21) Write each ratio in the previous exercise in its simplest form.

MULTIPLICATION BY RATIOS

$72 \times \frac{2}{9}$

$£1{\cdot}40 \times \frac{8}{7}$

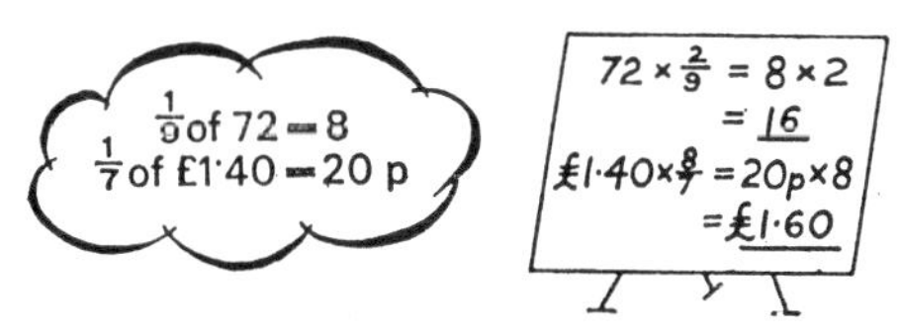

Find the value of:

107

(1) $35 \times \frac{1}{7}$ (2) $40 \times \frac{1}{4}$ (3) $45 \times \frac{2}{5}$ (4) $12 \times \frac{3}{1}$

(5) $72 \times \frac{3}{8}$ (6) $40 \times \frac{9}{2}$ (7) $120 \times \frac{3}{10}$ (8) $140 \times \frac{4}{7}$

(9) $135 \times \frac{5}{9}$ (10) $99 \times \frac{7}{3}$ (11) $20p \times \frac{2}{1}$ (12) $39 \text{ cm} \times \frac{1}{3}$

(13) $27 \text{ kg} \times \frac{10}{3}$ (14) $72p \times \frac{2}{9}$ (15) $49 \text{ litres} \times \frac{6}{7}$

(16) $£44 \times \frac{9}{4}$ (17) $£1{\cdot}60 \times \frac{7}{10}$ (18) $3{\cdot}05m \times \frac{6}{5}$

(19) $400m \times \frac{9}{10}$ (20) $£8{\cdot}76 \times \frac{4}{3}$

State if £24 is increased or if £24 is decreased in the next 4 examples:

(21) $£24 \times \frac{3}{8}$ (22) $£24 \times \frac{8}{3}$ (23) $£24 \times \frac{4}{6}$ (24) $£24 \times \frac{6}{4}$

(25) Find out when the value is increased and when it is decreased on multiplication by a ratio.

PROPORTION – RATIO METHOD

16 cakes cost 56p.
Find the cost of 10 cakes.

Use the ratio method of proportion to find the answers:

108

(1) 6 eggs cost 12p. What will 4 eggs cost?

(2) If 8 pens cost 72p then find the cost of 10 similar pens.

(3) On 12 litres of petrol a car can travel 96 km. How far will the car travel on 8 litres of petrol?

(4) If at a certain speed a car travels 90m in 5 seconds, how far will it travel in 30 seconds?

(5) Bill runs the same distance each day. In 1 week he runs 42 kilometres, how far does he run in 9 days?

(6) If 8 similar eggs weigh 960g, what is the weight of 6 of them?

(7) An American changed 12 dollars for £5. How many pounds would he get for 42 dollars?

(8) The ticket for a 20 km journey costs 45p. Find the cost of the ticket for a 32 km journey.

(9) I save the same amount each month. If I save £54 in a year, how many months will it take me to save £36?

(10) 45 grapefruit cost £1·80. Find the cost of 35 grapefruit.

NOTE: Quantities do not always change as simply as the previous exercise would suggest. *LOOK* at these questions:

If it takes 3 minutes to boil an egg, how long does it take to boil: (a) 2 eggs? (b) 3 eggs? (c) 4 eggs? (d) 5 eggs? (e) 500 eggs?

One of these questions cannot be answered. Why?

In Arithmetic always be prepared to use 'common sense'.

A road-roller can travel 210 metres in 27 seconds. How far can it travel in 45 seconds at the same speed?

Distance	Time
210m ⟷	27sec.
? ⟷	45sec.

Distance $= 210m \times \frac{45}{27} = 210m \times \frac{5}{3}$

Ans. = 350m

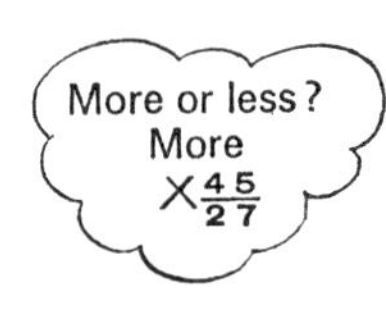

Find the answers to the following using the ratio method of proportion, where possible:

(1) 10 litres of petrol is needed for the car to travel 84 kilometres. How far can the car travel on 25 litres of petrol?

(2) If 3 men take 12 minutes to play a piece of music, how long will it take 4 men to play the same piece of music?

(3) If 6 women can sew 126 soft toys in a morning, how many soft toys can 10 women sew in the same time?

(4) Find the cost of 4 storage heaters if 14 cost £420.

(5) Standing on a hill 4 men can see a church spire 16 kilometres away. How far is the church spire from 2 of the men?

(6) The total wage bill for 16 men is £324. What is the wage bill for 20 men?

(7) In 3 hours an aeroplane flies 2 040 km. How far will it fly in 30 minutes at the same speed?

(8) Tom was 1·50 metres tall at 12 years of age. How tall will he be when he is 24 years old?

(9) 84 cm of scotch-tape can secure 18 parcels. What length of scotch-tape is required to secure 21 similar parcels?

(10) A factory employs 48 women to assemble 640 machines in a week. How many extra women must be employed to assemble 720 machines in a week?

Mixtures

DO AND CHECK

$8+2=10.$	Check: $10-2=8$ and $10-8=2$
$8-2=6.$	Check: $8=6+2.$
$8\times 2=16.$	Check: $16\div 2=8$ and $16\div 8=2.$
$8\div 2=4.$	Check: $8=4\times 2.$

110 Find the answers to the following. Write a check for each:

(1) $10+3$ (2) $5+12$ (3) $9-4$ (4) $29-14$
(5) $2{\cdot}9-1{\cdot}4$ (6) 6×9 (7) 6×19 (8) $24\div 3$
(9) $91\div 7$ (10) $9{\cdot}1\div 7$ (11) $3{\cdot}5\div 0{\cdot}7$ (12) $2{\cdot}4-1{\cdot}9$
(13) $5\times 8{\cdot}4$ (14) $2{\cdot}25\times 4$ (15) $5{\cdot}0\div 4$

111 Find $\square$. Write a check for each:

(1) $28\text{p}+6\text{p}=\square\,\text{p}$ (2) $49\text{p}+12\text{p}=£\square$
(3) $77\text{p}+36\tfrac{1}{2}\text{p}=£\square$ (4) $85\text{p}-37\text{p}=\square\,\text{p}$
(5) $90\tfrac{1}{2}\text{p}-71\text{p}=\square\,\text{p}$ (6) $50\text{p}-27\tfrac{1}{2}\text{p}=£\square$
(7) $54\text{p}\div 6=\square\,\text{p}$ (8) $28\text{p}\div 4=£\square$
(9) $15\tfrac{1}{2}\text{p}\times 8=£\square$ (10) £10·01 × 8 = £ $\square$

(11) 11) £10·01 = £ $\square$

(12) £10·01 − 2·02 = £ $\square$

(13) £10·01 + 9·09 = £ $\square$

112 Find the replacement value for each letter. Check your answers:

(1) $b+2{\cdot}5=6{\cdot}7$ (2) $7{\cdot}3+c=9{\cdot}9$
(3) $d-1{\cdot}3=4{\cdot}8$ (4) $12{\cdot}7-e=5{\cdot}4$
(5) $f\times 4=5{\cdot}6$ (6) $3\times g=0{\cdot}3$
(7) $h\div 5=10{\cdot}5$ (8) $10{\cdot}5\div j=2{\cdot}1$
(9) $\tfrac{1}{2}$ of $k=6{\cdot}8$ (10) $\tfrac{1}{3}$ of $x=5{\cdot}7$

$a+5{\cdot}1=9{\cdot}4$
$a=4{\cdot}3$
Check:
$4{\cdot}3+5{\cdot}1=9{\cdot}4$

Find and check the answers:

113

(1) There are 875 pupils on the roll of a school. 456 of these are boys. How many are girls?

(2) One plane ticket costs £14·65. Find the cost of 19 tickets.

(3) A bar of chocolate weighs 50 grammes. How many bars are there in a box weighing 1 kilogramme?

(4) An electric storage heater burns $2\frac{1}{4}$ units of electricity per hour. It is on for 8 hours each day. How many units does it burn in one week?

(5) A man's pay for a 40-hour week is £22·65. How much is he paid for one hour?

Progress Checks

PROGRESS CHECK 7

Find the answers:

(1) $24{\cdot}6 \times 3{\cdot}12$ (2) $30{\cdot}2 \times 7{\cdot}8$ (3) $26{\cdot}4 \div 0{\cdot}11$

(4) £$2{\cdot}45 \times 1{\cdot}25$ (5) What is 6% of £225?

(6) What is the ratio of 20p to £2·00?

(7) $482{\cdot}4 \div 0{\cdot}12$ (8) Express $\frac{1}{4}$ as a decimal

(9) Divide 1 080 by 72 (10) Subtract 48·23 from 71·19

PROGRESS CHECK 8

(1) Find 15% of 240 kg.

(2) What is the average weight of four boys weighing 58·2 kg, 60·4 kg, 59·7 kg and 63·7 kg respectively.

(3) A car travels at a constant speed of 60 km/hour. How far does it travel in 1·50 hours?

(4) Find the value of 10% of £12·50?

(5) Six men have an average weight of 91·5 kg. What is their total weight?

(6) A train has an average speed of 75 km/hour. A regular journey takes 4 hours. What distance does the train travel on this journey?

(7) An aircraft flies 6 006 kilometres in 3·5 hours. What was the average speed of the aircraft?

(8) If 9 books cost £7·02, what will 6 similar books cost?

(9) A shopkeeper pays a bill amounting to £465. He is allowed 10% off his bill for prompt payment. How much will he actually pay?

(10) Petrol costs $8\frac{1}{2}$p for one litre. How much will it cost a motorist to travel 300 km if his average petrol consumption is 15 km/litre?

PROGRESS CHECK 9

(1) Silk costs £2·54 per metre. What will be the cost of 8·25 metres of silk? (Approximate to the nearest 1p.)

(2) Copy and total this account:

6 dozen eggs at 30p per dozen
1·5 kg butter at 35p per kilogramme
2 litres of wine at £1·20 per litre
0·75 kg cheese at 48p per kilogramme

(3) A room is 6·5 metres long and 4·5 metres broad. If it is rectangular in shape, what is the area of the room?

(4) A tank holds 2 464 litres of oil. How many 2·2 litre cans may be filled from the full tank?

(5) 15 sacks have a total weight of 781·5 kg. What is the average weight of a sack?

(6) A man weighs 85 kg. His weight increases by 6%. What does the man then weigh?

(7) The petrol consumption of a car is 12·5 km/litre. 8·5 litres fill the petrol tank. How far does the car travel if the tank is full?

(8) An elevator can only carry a load of 800 kg. Six people, with an average weight of 70·5 kg, enter the lift. What weight can still be taken on the elevator?

(9) A colour television set is bought on hire purchase over a period of 5 years. If the set costs £310 and 5% service charge of the cost is added for each year, how much does the customer pay altogether?

(10) A man is paid £35·70 per week. 6% is deducted for superannuation. How much superannuation does he pay in one year (52 weeks)?